The Microtron

T0187982

THE PHYSICS AND TECHNOLOGY OF PARTICLE AND PHOTON BEAMS (formerly ACCELERATORS AND STORAGE RINGS)

a series of monographs edited by Swapan Chattopadhyay, Lawrence Berkeley Laboratory, California, USA

VOLUME 1
The Microtron, Sergey P. Kapitza and V.N. Melekhin

VOLUME 2
Collective Methods of Acceleration, N. Rostoker and M. Reiser

VOLUME 3
Recirculating Electron Accelerators, Roy E. Rand

VOLUME 4
Particle Accelerators and Their Uses, Waldemar Scharf

VOLUME 5
Theory of Resonance Linear Accelerators, I.M. Kapchinskiy

VOLUME 6
The Optics of Charged Particle Beams, David C. Carey

VOLUME 7
Getter and Getter-Ion Vacuum Pumps, Georgii L. Saksaganskii

VOLUME 8
Beam Dynamics: A New Attitude and Framework, Etienne Forest

VOLUME 9
Introduction to Radiation Acoustics, A.I. Kalinchenko, V.T. Lazurik and I.I. Zalyubovsky

VOLUME 10
The Microtron: Development and Applications, Yuri M. Tsipenyuk, edited by Sergey P. Kapitza

This book is part of a series. The publisher will accept continuation orders which may be cancelled at any time and which provide for automatic billing and shipping of each title in the series upon publication. Please write for details.

The Microtron

Development and Applications

Yuri M. Tsipenyuk
P.L. Kapitza Institute for Physical Problems, RAS, Moscow, Russia

Edited by

Sergey P. Kapitza
P.L. Kapitza Institute for Physical Problems, RAS, Moscow, Russia

CRC Press
Taylor & Francis Group
Boca Raton London New York

CRC Press is an imprint of the
Taylor & Francis Group, an **informa** business
A TAYLOR & FRANCIS BOOK

CRC Press
Taylor & Francis Group
6000 Broken Sound Parkway NW, Suite 300
Boca Raton, FL 33487-2742

First issued in paperback 2019

© 2002 by Taylor & Francis Group, LLC
CRC Press is an imprint of Taylor & Francis Group, an Informa business

No claim to original U.S. Government works

ISBN-13: 978-0-415-27238-4 (hbk)
ISBN-13: 978-0-367-39686-2 (pbk)

This book contains information obtained from authentic and highly regarded sources. Reasonable efforts have been made to publish reliable data and information, but the author and publisher cannot assume responsibility for the validity of all materials or the consequences of their use. The authors and publishers have attempted to trace the copyright holders of all material reproduced in this publication and apologize to copyright holders if permission to publish in this form has not been obtained. If any copyright material has not been acknowledged please write and let us know so we may rectify in any future reprint.

Except as permitted under U.S. Copyright Law, no part of this book may be reprinted, reproduced, transmitted, or utilized in any form by any electronic, mechanical, or other means, now known or hereafter invented, including photocopying, microfilming, and recording, or in any information storage or retrieval system, without written permission from the publishers.

For permission to photocopy or use material electronically from this work, please access www.copyright.com (http://www.copyright.com/) or contact the Copyright Clearance Center, Inc. (CCC), 222 Rosewood Drive, Danvers, MA 01923, 978-750-8400. CCC is a not-for-profit organization that provides licenses and registration for a variety of users. For organizations that have been granted a photocopy license by the CCC, a separate system of payment has been arranged.

Trademark Notice: Product or corporate names may be trademarks or registered trademarks, and are used only for identification and explanation without intent to infringe.

British Library Cataloguing in Publication Data
A catalogue record for this book is available from the British Library

Library of Congress Cataloging in Publication Data
A catalog record for this book has been requested

**Visit the Taylor & Francis Web site at
http://www.taylorandfrancis.com**

**and the CRC Press Web site at
http://www.crcpress.com**

CONTENTS

EDITOR'S PREFACE

This book on the microtron and its applications sums up the effort of more than 30 years of research and development of the microtron. Although the microtron, as an electron cyclotron was suggested in 1944 by Veksler in the very first of his papers on cyclic accelerators of fast particles, and the initial demonstration of the feasibility of this idea, it took many years for the first efficient and usable microtron to be developed. Most of this work was done at the Institute for Physical Problems of the Academy of Sciences in Moscow, and this research was first summarized in *The Microtron* written by V. Melekhin and myself. Published first in Russian and then in 1978 in English, it was the first volume of an international series of monographs on Accelerators and Storage Rings, edited by J. Blewett and F. Cole.

Since then many microtrons were built both in the Soviet Union and world wide, following the concepts and design worked out. This established the microtron as a useful and practical accelerator of electrons for energies up to 30 MeV with a beam power of a few kW. Although much more powerful linear accelerators for energies on many Gev were developed the microtron has a well defined and, I think unique ecological niche, that it occupies in the world of modern accelerators.

The kinematics of the accelerating process determine the well defined energy and momentum of the beam. Probably, this is the main advantage of the microtron as an accelerator. At the same time the intricate details of the accelerating fields limit the energy attainable in the classic design type of this accelerator. Certainly, by complicating the outlay and concept microtron accelerators have further evolved into complex and large machines, built for a special purpose and of a unique design. Thus the race-track microtron combines a linear accelerator with a magnet of some complexity in the construction of powerful reentrant accelerators.

Today with the advent and development of methods for calculating and optimizing electromagnetic fields and directly computing the motion of particles in an accelerator, the design of accelerators has to a great extent become a chapter in mathematical modelling. It is interesting to note, that the very first microtrons, developed by our group right from the beginning relied on detailed numerical calculations. This is because the microtron is really a very nonlinear device, and only by numerically calculating the orbits could a proper configuration of the accelerating cavity and injection be worked out. We were lucky to have access to the best computers then available and our computations were done, when precious computer time was given between calculations of space flights and the design of more ominous objects. It is sufficient to say that the fast memory available for coding directly in machine codes was 50 kilobits. Since then much has changed....

The book presents most of the work on the theory, design and operation of the microtron and accelerators based on its principle, when the phase changes in discontinuous steps from one orbit to the next. Dynamics of particles moving in a discrete phase space has now developed into a whole field of nonlinear mechanics, that may provide greater insight into the motion of electrons in microtrons. It is this peculiar motion of particles in a microtron that determines the well defined beam of this machine and makes it so suitable for injecting electrons into larger accelerators and storage rings. Finally, these properties of a microtron beam are well matched to the admittance of a free electron laser, providing a new field application of this versatile accelerator. The well defined beam also makes the microtron an ideal source for studying the passage of electrons through crystals.

Probably the most diverse applications of the microtron are its use as a source of secondary particles — bremsstrahlung, neutrons and positrons — and here the review of many fields of research will be of great interest to scientists coming from many branches of knowledge. At the same time these numerous experimental methods are of growing practical importance. From medical radiology to nondestructive testing the microtron has been shown to be well suited for these advanced and demanding uses, where again the well defined beam, simplicity and robustness make the microtron a very promising accelerator.

Much can be expected and has been done by the author in developing the uses of the microtron as a source of neutrons. Certainly, any reactor produces a much greater flux of neutrons, than a small electron accelerator. But now it has been shown that in many cases the microtron is not only a very potent source of neutrons for activation analysis, but as the result of the direct interaction of bremsstrahlung with nuclei it can lead to methods of activating practically all elements. This is a very significant development that should be properly evaluated in practical terms, when activation analysis is used for real work, rather than attractive, singular results, stunts, not staple methods for more mundane uses, that matter in the world of real things.

It may be suggested that the microtron is an ideal machine for a university, for teaching nuclear physics and developing applications in most diverse fields. At present a whole generation of research reactors have been or are to be phased out. Built many years ago at the dawn of the nuclear age they have gone through their life cycle. Many of these reactors are now rejected for environmental reasons and their image in the public mind is no longer attractive in terms of cost and inherent danger, perceived or real. An electron accelerator, the microtron, once switched off is practically harmless and on the other hand, as it has been shown in so many cases by Professor Tsipenyuk, a machine of infinite possibilities. I am sure that both those who are engaged in studying and designing accelerators, and the potential users will find this book to be a source of information and inspiration.

Sergey P. Kapitza

PREFACE

The history of the microtron now spans more than half a century from V.I. Veksler's proposal in 1944 and the construction of the first 'demonstration' machine in 1948 in Canada.

By the middle of the '60s the microtron had reached a state of development such that it could equal or surpass the performance of all other types of electron accelerators at energies up to approximately 30 MeV. The inherent simplicity, ease of construction, and low cost of the microtron resulted in its rapid development throughout the world for a wide range of scientific and technological applications. Some 20 years ago S.P. Kapitza and V.N. Melekhin published the book *The Microtron* in the series Accelerators and Storage Rings by Harwood Academic Publishers. The monograph covered theoretical and experimental results performed in those years and led to the widespread use and development of these accelerators. In 1984 the book *Recirculating Electron Accelerators* by R.E. Rand was published, and it mainly contains results on racetrack microtrons, which were beginning to be designed and built at that time.

Since then much work has been done at many laboratories and many accelerators of this type have been built for use as a versatile research tool in many fields. In fact, at present it is recognized, that powerful electron accelerators are a very useful source of electrons, gamma rays and neutrons for basic research and applications.

This book is a direct continuation of the monograph of S.P. Kapitza and V.N. Melekhin *The Microtron*, and it provides a systematic presentation of the main results in the investigation of the microtron and its applications made during the last 25 years. Recent experience at the Institute for Physical Problems as well as work at other places has been summed up. Besides research devoted to further development of the machine itself, a description of numerous scientific and technological applications on the microtrons are given.

The ubiquity of nuclear particles as a probe is now fully apparent over a wide variety of scientific fields and technology ranging from nuclear physics to solid state physics, chemistry, biology, metallurgy, geology, agriculture etc. Modern electron accelerators and the microtron in the first place as a powerful source of radiations generated by fast electrons determine the potential and multiple uses of these accelerators. For this reason, basics of the interaction of electrons, neutrons and gamma quanta with matter are also included.

The book also covers the use of electron accelerators in an analysis of structure and composition of different materials. The increasing demands for the purity of initial materials in modern science and technology impose claims of chemical impurities at the level of 10^{-12} part of the mass of the main matrix. Such a sensitivity can be provided only by nuclear methods. On the other hand, interest in trace elements, especially in biological and environmental systems, has been steadily increasing during the last few decades. Important fields of interest are animal, human and plant biology, food production, medicine and environmental studies. The interest in trace elements covers toxic elements as well as essential elements, some of which may also occur in dangerous concentrations. Another important aspect of trace elements in both organic and nonorganic matrices is that they sometimes display a specific pattern (often called *a fingerprint*), which may be typical for the origin or the history of a sample, for example, in archeology, tracing coins or even fissile materials.

A wide range of applications are conducted with the use of electron accelerators for non-destructive x-ray testing of large structures, e.g. nuclear reactors, vessels, chemical reactors. In this case a detailed and well documented study of the structure and of possible defects is produced. By using modern nuclear techniques an investigator can see the dynamics of the inner parts of the object. All these trends have led to numerous applications.

Great successes have been obtained in the application of specially designed microtrons for therapeutic use. At present the vast majority of cancer therapy facilities are based on electron accelerators. While circular medical microtrons are manufactured by corporation Agat in Moscow, the racetrack medical microtrons are produced by Scanditronix in Sweden. This is now a wide and well developed field of microtron activity. As to medical applications, most of the attention has now turned to the production of short-lived positron-active radioisotopes for positron emission tomography and diagnostical radioisotopes, in particular, iodine-123.

Conceptually the book consists of three main parts — new results on the microtron, secondary beams, and applications. I hope that the book will be of considerable value to many readers — post graduate researchers and students, practicing engineers, and scientists in many branches of research into the microtron and its applications. A review of the state-of-the-art in literature, and the many illustrations make the book of general interest to non-specialists in other areas of research. Also, those already established in the use of electron accelerators will find this book a useful summary of current knowledge.

Many results involved in the book were obtained with my friends and colleagues S.P. Kapitza, V.N. Melekhin, G.D. Bogomolov, A.R. Mirzoyan, B.S. Zakirov, V.E. Zhuchko, V.N. Samosyuk, B.A. Chapyzhnikov, V.I. Firsov, A.G. Belov, V.S. Potapchuk, Yu.T. Martynov, and A.V. Drobinin to whom I wish to express my deep gratitude for successful collaboration and numerous stimulating discussions of the problems under consideration.

Yuri M. Tsipenyuk

Part I. THE MICROTRON

Chapter 1

Basic Principles of a Microtron

1.1 Introduction

Veksler in his first paper [1.1] on phase stability in 1944 proposed a modification of the cyclotron for electrons, which is now called *the microtron*. This machine has a constant and homogeneous magnetic field and a constant accelerating rf voltage usually with wavelength $\lambda \sim 10\,\text{cm}$, i.e. it is the microwave band that gives the name of this machine. The same idea had already been mentioned in 1939 in the USA (by Alvarez, as it was cited in [1.2]) and in Japan in 1945 (by Itoh and Kobayashi [1.3]), but Veksler was the first one to publish. The first estimate of the phase and energy region where stable acceleration is possible was made in [1.3]. A few years after Veksler's original publication the race-track microtron was suggested by Schwinger (cf Schiff 1946 [1.4]).

The electron trajectory in a microtron is a system of circles, increasing in diameter, with a common tangent point where the accelerating cavity is placed — Fig. 1.1. The revolution period of electrons in the microtron after n transits across the accelerating cavity is

$$T_n = 2\pi E_n/ecB,\qquad(1.1)$$

where E_n is the total electron energy at n-th revolution and B is the magnetic field strength. Thus, the time required to complete one revolution is proportional to the total energy of the particle and an increase in period from one revolution to the next is directly proportional to the energy gain, i.e.

$$\Delta T = 2\pi\Delta E/ecB.\qquad(1.2)$$

1

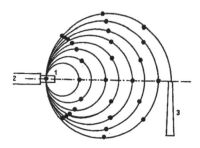

Figure 1.1. Circular microtron schematic layout (the dots represent electron bunches): 1 — accelerating cavity, 2 — waveguide, 3 — extracting channel.

If the energy gain per turn is adjusted to give an increase in period that is an integral multiple of the radio frequency, i.e. if the time of revolution in each orbit is one or more periods longer than in the previous orbit, the particles will return to the accelerating region at the same phase for each turn. The time taken for the first turn must also be an integral number of cycles of the radio frequency. These two conditions of synchronous motion of electrons at the microtron can be written in terms of the period of accelerating voltage T_0 in such a manner

$$\Delta E = \mu E_0, \quad T_1 = \nu T_0. \tag{1.3}$$

Here μ, ν are integer numbers.

It is more convenient to rewrite equation (1.2) in the form

$$\Delta E / E_0 = \nu B / B_0 = \nu \Omega \tag{1.4}$$

where $E_0 = mc^2$ is the electron rest energy, $\Omega = B/B_0$, $B_0 = 2\pi E_0 / e\lambda$ is the cyclotron magnetic field for wavelength λ of the accelerating field.

By means of the parameter Ω the conditions of resonance accelerating at the microtron can be written as

$$\Gamma_1 = E_1 / E_0 = \nu \Omega \tag{1.5}$$

$$\Delta \Gamma = \mu \Omega. \tag{1.6}$$

After n revolutions the relative energy of the synchronous electron will be

$$\Gamma_n = \Gamma_1 + n\nu\Omega. \tag{1.7}$$

Thus the total electron energy depends only on the number of orbits and the magnetic field intensity (on parameter Ω).

In modern microtrons the accelerating element is usually the cylindrical resonator in which the oscillation mode E_{010} is excited (Fig. 1.2). This kind of accelerating cavity was proposed and first made by S. Kapitza, V. Bykov and V. Melekhin [1.5]. The usage of such a cavity led to a great increase in the microtron's efficiency, and with such a cavity microtron beam power became comparable with the linac beam one.

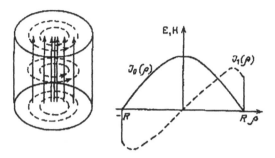

Figure 1.2. Electric (——) and magnetic (- - - -) field lines and corresponding field distributions in cylindrical cavity with E_{010} oscillation mode.

The electron injection from the cathode placed in one of the walls of the accelerating cavity was proposed by Melekhin. The initial electron orbits from the emitter are contained partially within the cavity itself as shown in Fig. 1.3 a,b. In other words, electrons gain energy inside the accelerating cavity before entering the first orbit, and, as it is considered below, this makes it possible to accelerate electrons with different energy gain per orbit, i.e. to change continuously the final electron energy by changing magnetic field strength and electric field in the cavity correspondingly.

Two modes of operation are possible — the 'first' and 'second' type (Fig. 1.3 a,b). Another injection scheme — from a small gun outside the toroidal cavity — was proposed in the early 60s in Sweden [1.6, 1.7], as it is shown in Fig. 1.3 c. Recently, an external gun was also used for a second type of acceleration in a rectangular cavity [1.8].

Consider how electrons being injected into the cavity from the cathode are caught into the acceleration (Fig. 1.3). The electron source is a small size LaB_6-emitter ($\sim 5\,mm^2$) heated by current in a direct or indirect way. The cathode is placed outside the symmetry axis.

The electrons are first accelerated inside the cavity along the trajectory shown in Fig. 1.3, and then leave the cavity near the

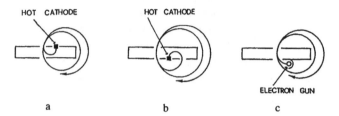

Figure 1.3. Conventional microtron injection schemes: by a hot cathode in the cavity wall: *a* — first type, *b* — second type; *c* — by an electron gun.

axis. Let $\delta E^{(1)}$ be the energy gain along this trajectory. Then

$$E_1 = E_0 + \delta E^{(1)} \tag{1.11}$$

and for $\nu = 1$ and $\mu = 2$ we obtain from relation (1.5)

$$\frac{1}{\mu}\frac{E_1}{E_0} = \frac{B}{B_0} = \frac{1}{2}\left(1 + \frac{\delta E^{(1)}}{E_0}\right). \tag{1.12}$$

One can see from the formula that the magnetic field in the microtron with this type of injection can be more or less that cyclotron field depending on the relation between the initial energy gain $\delta E^{(1)}$ and electron rest energy E_0. Thus one can continuously change the energy of the electron beam by changing the main magnetic field intensity and accordingly the microwave intensity in the cavity, i.e. the rf power coming into the cavity. The detailed calculations and experiments have shown that it is possible in practice to change continuously the electron energy in a fixed orbit approximately up to one and a half times.

Axial focusing in the circular microtron is assured only by the alternating electromagnetic field in the acceleration structure. According to the proposal of Melekhin [1.9], the shape of the entrance and exit apertures in the cavity has to be optimised for axial stability of the electron beam. The electrons are focused near the entrance hole and defocused near the exit hole by the transverse electric field, and they are defocused inside the resonator by the alternating magnetic field. Optimum axial stability is achieved by making the entrance aperture narrow in the axial direction, thus making axial focusing near the entrance hole, and by making the exit aperture rather large in the axial direction, thus weakening the axial defocusing near the exit hole.

1.2 Phase Stability

The energy spread of accelerated electrons is determined by the character of phase oscillations. Due to the principle of phase stability the accelerating process is self-correcting over a certain phase region, but not only for a particle that follows conditions (1.5) and (1.6).

The principle of phase stability was pointed out independently by Veksler (USSR) and McMillan (USA). According to this principle, radio-frequency electric fields can be applied in accelerators in such a way that particles undergoing acceleration will continue to experience this field at or near the correct phase of the alternating voltage cycle. The accelerating process is self-correcting over a certain phase region and can continue indefinitely without the particle falling out of phase with the field. It is conceivable that the peak voltage V_0 across the accelerating gap could be so chosen that the energy eV_0 gained by an ion in crossing at the peak would be just that needed to raise the energy to the higher value of the total energy of synchronous particle E_s required during the next turn. But the number of such successfully accelerated ions would be extremely small. Particles in an accelerator do not advance entirely side by side, but are also spread out into a column of appreciable azimuthal length. One might then expect that if the front end of such a column reached the accelerating gap of maximum voltage V_0 all later ions would receive less than the necessary energy gain. Consequently, the column would lengthen and many of the ions would undergo deceleration because of arrival at the gap when the voltage was directed the wrong way.

There is a natural solution to this difficulty as pointed out by Veksler and McMillan, provided certain easy requirements are fulfilled. The requirements are:

(i) the peak voltage across the accelerating gap must be somewhat greater ·than would otherwise be needed;

(ii) the particle orbits must be such that a change in energy and momentum must be accompanied by a change in the period of revolution.

To demonstrate the way in which phase stability acts, let us assume the magnetic field and accelerating voltage are chosen such that a synchronous electron passes the cavity at an instant of time that corresponds to phase φ_0 (Fig. 1.4).

Now consider an electron that reaches the accelerating gap earlier at phase $\varphi < \varphi_0$. It now experiences larger energy gain than the synchronous electron. This makes its period of revolution exceed the period of the synchronous electron. So the gap is next encountered one turn later at a phase closer to φ_0. The energy is then raised again, though by a lesser amount, and the subsequent phases shift further in the same direction. Reasoning similarly, we come to the conclusion that if electrons cross the accelerating gap too late they acquire less energy and begin to overtake the synchronous particle. It just means that during the process of acceleration electrons execute phase oscillations about the synchronous value.

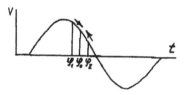

Figure 1.4. Time-dependence of the electric field in an accelerating cavity. Arrows show the direction of phase motion of non-equilibrium particles that leads to phase stability during acceleration.

The phase change between turns can be described by difference equations. Phase to energy and energy to phase relationships are of interest in the analysis of phase motion. The time difference between the two successive orbits equals $\nu\tau$. So, the time increase corresponds linearly to the increase in energy. Introducing the energy offset in the nth orbit, $\delta E_n = E_n - E_{ns}$, from the energy of a synchronous electron which has the same energy gain, ΔE, between all orbits, the difference in increased time becomes:

$$\delta T_n = \Delta T \frac{\delta E_n}{\Delta E} = \nu\tau \frac{\delta E_n}{\Delta E}. \tag{1.13}$$

With a phase error, $\delta\varphi_n$, from the synchronous phase φ_s, the change in phase error due to the excess energy, δE_n, in the nth orbit is:

$$\delta\varphi_{n+1} - \delta\varphi_n = 2\pi\nu \frac{\delta E_n}{\Delta E}. \tag{1.14}$$

The phase is defined as positive on the falling side of the accelerating wave such that $\Delta E = eV_0 \cos\varphi_s$, where V_0 is the amplitude of the accelerating voltage. Note, that V_0 stands for the efficient accelerating voltage taking into account transit time effects.

The increment in energy excess which the electron gets as it passes $(n+1)$ through the accelerating cavity is:

$$E_{n+1} - \delta E_n = eV_{\text{eff}} \cos(\varphi_s + \delta\varphi_{n+l} - eV_{\text{eff}} \cos\varphi_s \simeq$$

$$eV_{\text{eff}} \delta\varphi_{n+1} \sin\varphi_s = -\Delta E \, \delta E \delta\varphi_{n+1}\tan\varphi_s \,. \qquad (1.15)$$

Equations (1.14) and (1.15) may be described by the matrix equation:

$$\begin{bmatrix} \delta\varphi_{n+1} \\ \frac{\delta E_{n+1}}{\Delta T} \end{bmatrix} = \begin{bmatrix} 1 & 2\pi\nu \\ -\tan\varphi_s & 1 - 2\pi\nu\tan\varphi_s \end{bmatrix} \begin{bmatrix} \delta\varphi_n \\ \frac{\delta E_n}{\Delta E} \end{bmatrix} . \qquad (1.16)$$

Since the determinant of the matrix is unity, the stability condition is:

$$\left[\frac{1}{2}\mathrm{Tr}(M)\right]^2 < 1 \qquad (1.17)$$

which means that:

$$(1 - \pi\nu\tan\varphi_s)^2 \quad \Longleftrightarrow \quad 0 < \pi\nu\tan\varphi_s < 2. \qquad (1.18)$$

For $\nu = 1$, which gives the widest phase stable region, the synchronous phase can be within the range 0 to 32.5°. For the first orbit completely encircling the resonator the parameter μ has to equal 2.

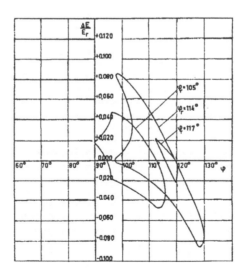

Figure 1.5. Microtron phase area for various resonant phase angles φ_r (from [1.7]).

Phase stable acceleration occurs for phase angles shortly after the peak of the resonator voltage. With the introduction of a few simplifying assumptions, it is possible to estimate the angular region of phase stability. The resonator electric field is assumed to be homogeneous and the gap width so short that the presence of the dc magnetic field within the gap may be neglected. For all transits except the first, the electron velocity is assumed to be constant and equal to light velocity c.

For an infinitely thin accelerating gap phase motion was initially treated in detail by Kolomensky [1.10]. The electron phase and energy error, as found numerically, can be represented in diagrams as shown in Fig. 1.5. If the initial phase is chosen within certain limits, the values of φ and ΔE for successive transits form a closed curve. An electron with phase and energy error within the boundary can be stably accelerated.

In fact, the phase trajectory for a finite thickness accelerating cavity differs from the idealised case shown in Fig. 1.5, but qualitatively the character of phase motion is the same.

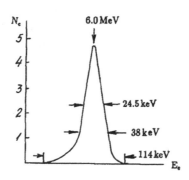

Figure 1.6. Energy spectrum of electrons in the racetrack microtron [1.12]; the width of the spectrum is pointed at the levels 0.5, 0.75 and 0.95.

The phase stability conditions imply that the output energy is restricted within a finite and reasonably small envelope. An important peculiarity of the microtron follows from the equations of phase oscillations: the absolute electron energy spread is proportional to the dimensionless parameter Ω, but doesnot depend on the number of orbits, i.e. on the electron energy. The total energy interval of phase motion in the first accelerating type accounts for around $0.1mc^2 \simeq 50$ keV. Henderson *et al.* [1.11] found an energy spread of 50 keV in a 6.8 MeV beam, and Wernholm [1.7] measured an energy spread of $\pm1\%$ in a 5.94 MeV beam. These values of energy spread were obtained from conventional (circular) microtrons. The energy spectrum of the beam of a four-sector racetrack microtron has been measured by Sells *et al.* [1.12] and is shown in Fig. 1.6. All these results are in good agreement with the theoretical calculations.

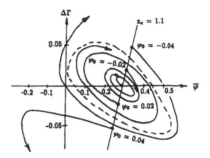

Figure 1.7. Phase motion of electrons in a microtron using I type of acceleration; x_c is the coordinate of the cathode (in reduced units), φ_0 is the injection phase.

Numerous calculations have been performed to find the region of stable acceleration depending on different microtron parameters. Figure 1.7 illustrates the phase motion of electrons in the first type of acceleration. One can see that there is an ellipse on $(\Delta\Gamma, \overline{\varphi})$ plane (dashed line) that restricts the region of stable acceleration process.

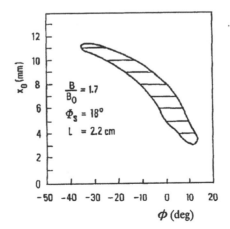

Figure 1.8. Extension of capture phase region versus the distance (x_0) of the emitting point from the axis of the resonator (from [1.13]).

From a practical point of view it is more convenient to present the results of the numerical calculations as the dependence of emitting point versus capture phase (Fig. 1.8), and the capture efficiency versus emission position (Fig. 1.9).

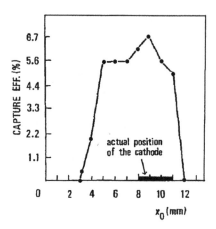

Figure 1.9. Capture efficiency versus emission position x_0 (from [1.13]).

In the following chapters we consider the development of numerical calculations of stable acceleration in a circular microtron, experimental improvements of the accelerating process, new injection schemes, and modern construction of accelerators.

Chapter 2

Circular Microtron

2.1 Construction

The microtron is usually a pulsed accelerator as it is necessary to feed to the accelerating cavity a rather high microwave power — about 300-400 kW — but there is nothing restricting the machine from operating in continuous wave regime. The conventional microtron operates with the repetition rate 100-1000 Hz and a pulse length of some microseconds; the average power of the accelerated beam is of some kW and energy up to 30 MeV. The size and weight of the microtron are comparatively small — the diameter of the magnet is about 1-1.5 m, the weight $\sim$ 1500 kg. Figure 2.1 is a basic diagram of a microtron.

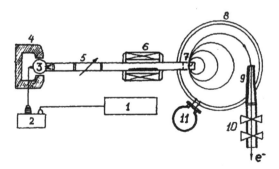

Figure 2.1. The schematic diagram of a microtron: 1 — modulator, 2 — pulse transformer, 3 — magnetron with its magnet 4, 5 — phase shifter, 6 — ferrite insulator, 8 — vacuum chamber, 9 — extractor, 10 — quadrupole lenses, 11 — vacuum pump

11

2.1.1 Magnet

The essential features required of the vacuum chamber of a microtron are the following: it has to be mechanically rigid, nonmagnetic, it has to provide high vacuum, and it is desirable for it to be technologically simple to manufacture. As a rule, the vacuum chamber is made from stainless steel or brass. The use of stainless steel has one strong disadvantage, is that it can become magnetic after welding.

On the other hand, it is possible to refuse from a separate vacuum chamber and to extend its vacuum function to magnetic core. Such a construction was performed in JINR (Dubna) [2.1] and the cross-section of this electromagnet is shown in Fig. 2.2. The coils are placed in the vacuum and are made from rectangular coaxial cable with an internal channel for water-cooling. To simplify the correction of the magnetic field the shims can be dismantled into rings of necessary shape.

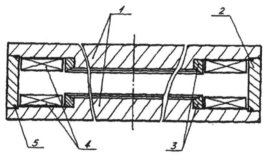

Figure 2.2. Construction of the microtron electromagnet at JINR (Dubna): 1 — magnet pole, 2 — magnetic core, 3 — circular shims, 4 — coils, 5 — vacuum seating.

Usually, the magnetic poles are not parallel after manufacturing, and this is due to wearing of the tool during the machining, as illustrated in Fig. 2.3 (curve 1). It is necessary to keep in mind that the poles tend to be pulled out of line by atmospheric pressure. This tendency is also shown in Fig. 2.3.

The same problem was solved in the Wisconsin University microtron [2.2] by another way. The microtron serves as an injector for electron storage ring Tantalus. In this machine, having 34 orbits (final electron energy 44 MeV), the pole diameter equals 48 cm and even if the poles were 4.5 in. thick they would deflect approximately 0.002 in. under the air load because of their large diameter. To avoid the difficulty of possible magnetic field inhomogeneity the pole pieces were split in two. This may be

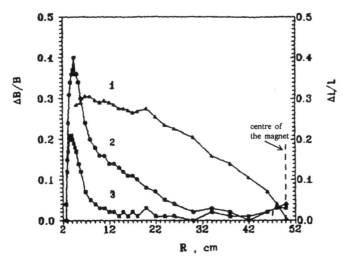

Figure 2.3. Distribution of the magnetic field at the JINR microtron: 1 — parallelism of magnetic poles in %, 2 – magnetic field at atmospheric pressure, 3 — magnetic field with vacuum inside the volume.

seen in Fig. 2.4. The top and bottom pole pieces carry the air load while the inner pole pieces, which will be in the vacuum chamber, will have to support only their own weight. The gap between the two pole pieces also serves as a field homogenizer.

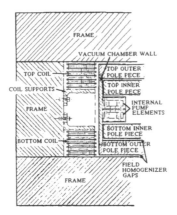

Figure 2.4. Wisconsin microtron cross section.

It is also seen from Fig. 2.4 that internal sputter-ion pumps operating in the magnet fringe field are used. Spacing of ion pumps inside of vacuum chamber is now made in many microtrons. Special additional coils are used to correct magnetic field irregularities. Suppose we have a radial average field $\overline{B_r}$ acting on all orbits. The vertical displacement on the nth orbit will be

$$\Delta z_n = 2\pi \overline{B_r} \lambda (n+1)/B . \tag{2.1}$$

The total displacement for electrons of energy E_0 is obtained by the sum of all orbits

$$\Delta z \simeq \frac{N^2}{2} \overline{B_r} \frac{\lambda^2 c}{E_0 \Delta \gamma} . \tag{2.2}$$

The maximum tolerable displacement is governed by the half height of the cavity aperture $h/2$. For example, this means that for $N = 20$, $h/2 = 4\,\text{mm}$, and $\lambda = 10\,\text{cm}$ we find that the allowable value of the radial component of a guided magnetic field equals $\sim 5 \times 10^{-2}\,\text{G}$.

In order to correct this type of field irregularities, which lead to displacement of the orbits in the vertical direction, in the Italian microtron [2.3] a couple of coils wound around the return core of the magnet were used (see Fig. 2.5). These coils are excited in opposition to each other in order to leave the net B_z field unchanged but at the same time correct the radial component in the plane of symmetry B_r.

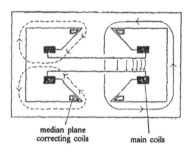

median plane
correcting coils   main coils

Figure 2.5. Median plane correcting coils: – – – – B field lines from the correcting coils, —— B field lines from the main coils.

In the case of slight misalignment of the poles or iron defects we deal with left-right asymmetry of the magnetic field. This type of asymmetry leads to shifting of the centres of the orbits on the symmetry plane towards or away from the cavity, depending on the sign of the asymmetry.

If we call ΔB the field difference between the right and the left side of the magnet, the nth displacement is equal to

$$\Delta x_n \simeq 2n\lambda \Delta B/B \tag{2.3}$$

and the total displacement will be

$$\Delta x \simeq N^2 \lambda \Delta B / B \qquad (2.4)$$

which, with a maximum tolerable shift of 5 mm, leads for $N = 20$ to a maximum $\Delta B = 0.2$ G.

This effect is corrected by the use of a pair of coils placed at the opposite sides of the resonator and generating opposite fields. The role of these coils in tuning the orbit position and intensification of vertical focusing will be considered below.

In a microtron magnet the magnetic field is only homogeneous in its central region, and the smaller the size of the magnet, the smaller the relative part of this region. This point is most sufficient in a microtron for small final energy ($\sim 5 - 7$ MeV). For instance, in a 10-cm wavelength microtron with 30 orbits the working field is about 80% of the total pole-piece surface while for a 10 orbit microtron it is $\sim 60\%$.

Non-uniformities of the magnetic field disturb electron orbits and lead to losses of accelerated electrons. The most critical inhomogeneities for the process of electron acceleration are those which are asymmetric relative to the common diameter of orbits. The allowable relative inhomogeneity of this type at the last Nth orbit, as it follows from (2.2) and (2.4), is proportional to $1/N^2$. Commonly just this relation is used as the criterion for necessary magnetic field inhomogeneity.

On the other hand, the requirements which are imposed on other types of non-uniformity can have significantly lower tolerances. For instance, in a magnetic field which is linearly or quadratically decreased from both sides of the common diameter, the non-uniformity can reach some percent.

It was demonstrated experimentally that it is actually possible to expand the working field of a microtron by use of the non-uniform magnetic field at the edge of the magnetic pole [2.4]. The authors studied acceleration with extraction of the resonator from the uniform-magnetic-field region. The working magnetic field can be expanded by shifting the sharp drop of the magnetic field induction closer to the edge of the pole. This can be done by changing the shape of the pole's edge. Figure 2.6 illustrates how the profile of the edge of the magnetic pole changes the radial distribution of the magnetic field.

It was confirmed experimentally that the capture coefficient in the microtron with the resonator shifted by two orbits closer to the edge of the magnetic pole with the pole-piece shape 3 (see Fig. 2.6) is approximately the same as in acceleration in a uniform

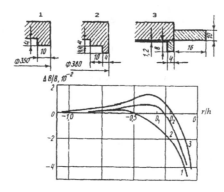

Figure 2.6. Magnetic pole-piece shapes and magnetic field distributions correspondingly, r is the distance from the edge of the magnetic pole, h is the height of the gap between poles.

magnetic field. Small shifts in the position of the last orbit and an increase of the horizontal size of the accelerated beam were eliminated by using correction magnetic coils.

Another way of expanding the working region of the magnetic field was proposed by Melekhin [2.5]. The main idea of this proposal is to make special radial ridges in magnetic core, as it is shown in Fig. 2.7. In such a configuration of magnetic poles the distance l from the edge of these ridges to the reverse magnetic core becomes less than the gap h between the magnetic poles.

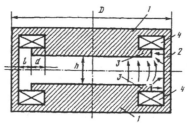

Figure 2.7. Cross section of the electromagnet: 1 — poles, 2 — reverse magnetic core, 3 — radial ridges, 4 — coils. Magnetic field lines are shown by arrows.

Such a geometry of magnetic poles results in radial components of the magnetic field in the gap between ridges and reverse magnetic core, and the strength of this radial field is higher than the guiding magnetic field between the poles. On the other hand, such configuration of magnetic poles also has to expand the uniform-magnetic-field region. This idea is used, in particular, in the 8 MeV microtron-injector for a 200-600 MeV compact electron synchrotron [2.6].

In designing the microtron, much attention is now paid to its compactness together with high reliability and stability of parameters and to high technological ability for prophylactic control. Great experience in designing and manufacturing of accelerators, and the use of modern technology allows a very compact microtron of variable energy to be built, which could be used for many applications, for instance, in free electron lasers of far infrared range, for electron synchrotrons, and also in medical installations for electron and γ-therapy. Two microtrons, in which the ideas discussed are used, are considered below.

As an example, let consider the microtron of variable energy $2.3 \div 22.4$ MeV which is under development in Budker Institute of Nuclear Physics (Novosibirsk) [2.7]. The accelerator vacuum volume in this machine is bounded by the magnetic core and poles which are extended by a thin crosspiece up to the reverse core (Fig. 2.8).

The metallic (indium) sealing is used, ensuring the vacuum is clear (there are no organic admixtures in it) and increasing the cathode lifetime.

To get high homogeneity of the magnetic field there are two correction coils inside each pole and a shim at the outer radius of the poles. The shim location and shape as well as the correction coil location and currents optimised using numerical computer calculation. The picture of the magnetic flux and plot of magnetic field flat top from this calculation are also presented in Fig. 2.8.

2.1.2 Microwave System

The microwave system of the microtron usually consists of a magnetron (klystron), ferrite insulator or circulator, phase changer, and microwave window which separates the vacuum in the accelerating chamber from the air in the waveguide. Magnetrons have generally been preferred over klystrons because of their high efficiency and lower anode voltage requirement. The coupling between the cavity and the waveguide is accomplished through the coupling hole the dimension of which has to be chosen in such a way as to provide the highest efficiency of microwave power transformation into the accelerating electron beam.

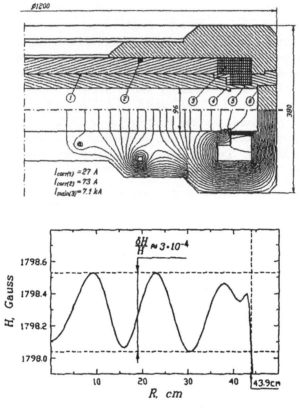

Figure 2.8. *Upper:* Cross-section of the magnetic vacuum system in regular part and the field flux picture: 1, 2 — correction coils, 3 — main winding, 4 — shim, 5 — ferromagnetic crosspiece, 6 — vacuum sealing. *Below:* Magnetic field in median plane.

2.1.2.1 Energy Relations

Let the relative fraction of the direct losses of rf power in the ferrite insulator equal ξ and assume the magnetron supplies the power P_m. It means that the power incident on the cavity from the waveguide equals $P_w = P_m(1 - \xi)$. In such a situation the power P_a absorbed by the cavity is

$$P_a = \frac{4\beta}{(1 + \beta)^2} P_w.\qquad(2.5)$$

Here the parameter β characterises the coupling of the cavity with the power-supplying waveguide. The absorbed power is maximal at $\beta = 1$, and, certainly, there is no reflecting power from the cavity in this case.

The connection between the coupling coefficient β and the coefficient of the amplitude of the standing wave k, that is experimentally measured, is well known:

$$k = \beta \quad \text{if} \quad \beta > 1 \quad \text{and} \quad k = 1/\beta \quad \text{if} \quad \beta < 1. \tag{2.6}$$

Let $\beta = \beta_0$ in an unloaded cavity where the absorbed power is practically equal to its ohmic losses P_o. The loading power of the microtron cavity

$$P_1 = P_e + P_{nr} \tag{2.7}$$

consists of the power of the accelerated electron beam $P_e = IE_e$ with the final energy E_e at the average macropulse current I and the power P_{nr} consumed by non-resonant electrons. Actually, $P_w = P_o + P_e + P_{nr}$.

For the cavity loaded with an accelerated electron beam the equation (2.5) becomes

$$P_1 + P_o = \frac{4\beta}{(1 + \beta)^2} P_w, \quad \beta = \frac{\beta_0}{1 + P_1/P_o}, \tag{2.5a}$$

where β is the loaded coupling coefficient. From these relations we can obtain that

$$P_w/P_o = (1 + \beta_0)^2/4\beta_0 = (1 + \beta_0 + P_1/P_o)^2/4\beta_0. \tag{2.8}$$

The loading power can be expressed from (2.8) in the form

$$P_1 = \left[2\sqrt{\beta_0 P_w P_o} - (1 + \beta_0)P_o \right]. \tag{2.9}$$

This relation determines the power load of the cavity which is allowed by the given microwave parameters of the microtron.

The maximum allowed microwave power P_{max} which can be transferred to the accelerated electron beam is $P_{max} = P_m - P_o$, when the reflecting power from the cavity does not exist $(\beta = 1)$ and the power loss in the ferrite insulator is relatively small $(\xi \simeq 0)$. In this case the optimal coupling coefficient

$$\beta_0^{opt}(P_m) = P_m/P_o \tag{2.10}$$

is only determined by the power which is available from the rf generator and ohmic losses in the cavity.

2.1.2.2   Coupling of Resonator with a Waveguide

In order to understand qualitatively the excitation of an accelerating cavity through the hole in one of its walls from a waveguide, we consider first a more simple case of excitation of a cavity through a small coupling hole. This problem was, in particular, solved in detail by Wainstein [2.8].

We will consider only the case of a commonly used magnetic excitation. Consider first an ideally conductive plane with a hole the dimension of which is small relative to the rf wavelength λ. At the left we have electromagnetic fields E_1, H_1 undisturbed by the hole, and at the right — E_2, H_2. In the right half-space the radiative field is governed by the electric and magnetic dipole moments of the hole

$$d_2 = \alpha(E_2 - E_1), \quad m_2 = \kappa(H_2 - H_1). \tag{2.11}$$

In the left half-space the radiative field is determined by the moments

$$d_1 = -d_2, \quad m_1 = -m_2. \tag{2.12}$$

The polarization factors α and κ are the solution of a static problem and certainly depend on the shape of the hole. For a circular hole of radius $r \ll \lambda$ in an infinitely thin wall these coefficients are equal to

$$\alpha = r^3/3\pi, \quad \kappa = 2r^3/3\pi. \tag{2.13}$$

Now we can turn our attention to excitation of the microtron accelerating cavity through a finite hole in its wall by a waveguide. For simplicity we will make three suppositions:

1) in the waveguide an external rf power supply excites only one waveguide mode which transfers a power P_0;

2) the metallic wall between a waveguide and a cavity is infinitely thin;

3) excitation is performed through a small circular hole of a radius $r \ll \lambda$ and only magnetic type of excitation occurs.

The electromagnetic field of the waveguide near the hole can be approximated as the sum of the undisturbed field in the waveguide with a plunger and the field of an elementary magnetic dipole being placed in the centre of a coupling hole. If there is no normal component of electric field in the hole and there are undisturbed magnetic fields H_1 in the waveguide and H_2 in the cavity, then the quasistatic magnetic moment of the window is

$$m = \kappa(H_1 - H_2), \quad \kappa = 2r^3/3\pi. \tag{2.14}$$

Here κ is the coefficient of static magnetic polarization of the hole.

The amplitude of electromagnetic field in the cavity is proportional to the value of the dipole magnetic moment of the coupling element, and therefore for a small hole of the radius r the electromagnetic power transmitted through a hole into the cavity is proportional to the value of the magnetic moment squared. Thus, the power in a cavity is proportional to the sixth power of the radius of the hole.

If the wall between a cavity and waveguide is of finite thickness d we can consider a circular hole of radius r in it as a section of a circular waveguide in which there are only attenuated waves. If the electromagnetic field in the plane of the hole has only tangential magnetic field (this is practically the case in a microtron) then in the hole mainly H_{11} waveguide mode is excited, and consequently at the opposite side of the hole (in the cavity) the amplitude of the $H11$-wave is attenuated by $\exp(-1.841d/r)$ times.

Therefore, for a hole of the finite thickness the polarization coefficient has to be written in the form

$$\kappa_m = -\frac{2r^3}{3\pi}e^{-1.841d/r} . \qquad (2.15)$$

The dependence of the coupling coefficient β for cylindrical cavity having a rectangular hole height $11\,\mathrm{mm}$ on its width a was measured by different authors. Figure 2.9 shows the results of measurement performed by A.B. Manenkov. The results agree well with the expression

$$\beta = 1.22 \times 10^{-9} \times a^{6.6} , \qquad (2.16)$$

which qualitatively follows the considerations given above.

2.1.2.3 Automatic Frequency Control System

A stable acceleration process in a microtron is ensured by coincidence of the frequencies of the microwave oscillator and the accelerating cavity. Theoretical estimates and experimental investigations show that in order to maintain stability of the accelerated electron current with an accuracy of 10% the relative frequency drift of the oscillator must not exceed $\pm 0.01\%$ for a loaded-cavity $Q_l \sim 1000$. In magnetron oscillators, however, frequency drifts by several tenths of a percent are commonly observed when the operating mode changes. Moreover, the natural frequency of the accelerating cavity changes due to its heating.

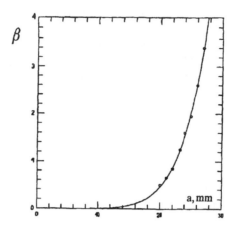

Figure 2.9. Coupling coefficient of the cavity with a waveguide through the hole 11 mm height and different widths.

That is why a system for automatic frequency control of the accelerating cavity was developed [2.9]. The system eliminates the mutual detuning of the oscillator and cavity when the microtron operates. This system simplifies the control of a microtron during the start-up process and during changes in its operating mode.

Figure 2.10 shows the block diagram of the automatic frequency control system of the accelerating cavity of a microtron. The basic element of the automatic frequency control system is a mismatch sensor with phase detection. The sensor uses the dependence of the phase shift between the field intensity in the cavity and the field intensity of the incident wave in the waveguide channel on the oscillator frequency. The mismatch sensor consists of a ring bridge 6 based on a strip line, two aperiodic coaxial detector chambers 7 and 8, phase changer 5, and the accelerating cavity 4. At the oscillator frequency, which is equal to the natural frequency of the cavity, a phase shift of 90° is established between the signals applied to the ring bridge 6 from the cavity 4 and the waveguide channel 2 by means of the phase changer 5. For detuning of the cavity relative to the oscillator frequency the phase difference between the input signals of the ring bridge will differ from 90°. Under these conditions a difference signal appears at the output of the comparison circuit 14 and causes the actuating element 12 to tune the cavity 4 to the frequency of the magnetron oscillator 1.

The bridge 6 is made of a strip line in the form of a ring $3/2\lambda_n$ long with branches spaced $\lambda_n/4$ apart (λ_n is the wavelength

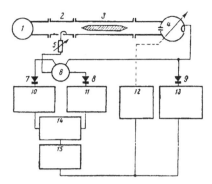

Figure 2.10. Block diagram of the automatic frequency control system: 1 — microwave oscillator (magnetron), 2 — waveguide section with a coupling loop, 3 — ferrite gate, 4 — accelerating cavity, 5 — phase changer, 6 — ring bridge, 7-9 — detector chambers, 10,11 — video amplifiers, 12 — actuating element, 13 — amplitude discriminator, 14 — comparison circuit, 15 — power amplifier.

of the strip line). The strip line is made from brass foil 0.1 mm thick which is placed between two dielectric plates 2 mm thick having a relative dielectric constant of 2.45. The dielectric plates are covered on the outside with conducting sheets.

The output stage of the power amplifier 15 includes two thyristors and a differential output transformer. The voltage taken from the output winding of the transformer is applied to the control winding of a motor. The latter, combined with a reducer and a tachometer-generator, forms an actuating element which tunes the frequency of the accelerating cavity by deforming one of the cavity walls.

The operation of the automatic frequency control system was checked experimentally on a 25 MeV microtron. The dependence of the average current of accelerated electrons on magnetron power was determined. This dependence was obtained both with the automatic frequency control in operation and without it. In the latter case the cavity was tuned to the magnetron frequency manually at the minimum power level; no tuning of the cavity was carried out for a further increase in oscillator power. The results of these experiments are shown in Fig. 2.11. The difference in the shape of curve *b* from that of curve *a* in Fig. 2.11 is connected with the fact that when the magnetron power increases, the frequency changes and resonance in the system is violated; the accelerated current undergoes practically no increase. Note that in both cases the magnitude of the emission current corresponds to the maximum beam current at the given oscillator power.

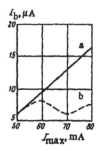

Figure 2.11. Dependence of the average current of accelerated electrons on the magnitude of the magnetron anode current: *a* — with automatic frequency control, *b* — without automatic frequency control.

The automatic frequency control system normally operates in a frequency band ±0.1%, which is determined by the characteristic of the mismatch sensor. For a larger detuning between the oscillator and cavity the automatic frequency control system operates in a search mode. The signal from the accelerating cavity is used as the signal for switching the automatic frequency control system from the search mode to the follow-up mode.

The automatic frequency control system considered completely eliminates the detuning between the oscillator and cavity and allows a substantial simplification in microtron control during its start-up and operation. It should also be noted that when the microtron operated with a tunable magnetron the automatic frequency control system considered provided tuning of the magnetron to the cavity frequency.

2.1.2.4 Platinotron as a Microwave Power Supply

Commonly in microtrons, magnetrons and klystrons are used with a pulse length of some microseconds. On the other hand, there are modern high power rf generators that can provide longer pulse length, which can be some tens of microseconds. In such a case we actually have quasi-continuous mode of operation. The reality of operation of such a microtron was demonstrated by Stepanchuk with colleagues who used a platinotron as an rf generator [2.10].

The platinotron was operated in an autogenerating regime with a positive feedback loop from an accelerating cavity. The main feature of the resonator is that it has two holes in its walls, as it is shown in Fig. 2.12. The main hole serves for power supplying from the generator through the waveguide, and the additional hole provides a feedback signal.

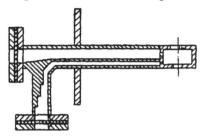

Figure 2.12. Accelerating cavity with two holes for power supply and feedback signal; the waveguide which provides feedback signal, has variable cross section with a Chebyshev-like transformer.

As it was demonstrated in [2.10], stable operation of a microtron with a platinotron is possible and even with a $40\,\mu s$ pulse length there is no arcing in the resonator and a usual LaB_6 cathode provides an injection and acceleration of such a long pulse.

2.1.3 Extraction of Electron Beam and Its Injection to Synchrotrons

2.1.3.1 Extraction of Accelerated Electron Beam

For a microtron operated at 10cm wavelength the separation between the consecutive orbits is $\lambda/\pi \simeq 3$cm. Plenty of space is therefore available to extract the beam. The simplest beam extractor is a soft iron tube arranged to catch the beam tangentially and carry it out of the magnetic field region through its screened inside. Even if it catches the beam where the orbit separation is largest, such an extractor will, however, considerably lower the magnetic flux density in a region extending over the two previous orbits. The effect of this distortion is to push the centres of curvature of the previous orbits towards the resonator. The inhomogeneity of the field also changes the beam shape. Therefore, special shims are needed to compensate the field distortion by extraction channel.

The size and shape of an extraction channel is designed according to the divergence of the accelerated electron beam. The radial angular divergence of a beam $\Delta\theta$ at the point of the orbit of radius R, which is on the common diameter one can estimate by radial size of the beam in the perpendicular direction

$$\Delta\theta_r = \Delta y_r / R. \tag{2.17}$$

As a rule, the radial size of the beam Δ_r, is about $5\,\mathrm{mm}$, and consequently, for instance, for the 20th orbit in a $10\,\mathrm{cm}$ wavelength microtron ($R = 30\,\mathrm{cm}$) the radial angular divergence $\Delta\theta_r \simeq 2 \times 10^{-2}$.

Vertical angular divergence is governed by vertical oscillation of a beam and can be estimated in the following way

$$\Delta\theta_z \leq \frac{\sqrt{2}}{n\pi} Z_{\max}, \tag{2.18}$$

where n is the number of the orbit, $Z_{\max} = (2\pi/\lambda)Z$ is the maximal amplitude of vertical oscillations which is determined by size $Z \simeq 0.25$ of the aperture of the accelerating resonator. Therefore,

$$\Delta\theta_{z\max} \simeq 0.11/n. \tag{2.19}$$

So, vertical angular divergence is about 5 times less than radial divergence, and the latter determines the size of the extraction channel.

The usual geometry of the extraction channel with compensation rods is shown in Fig. 2.13. This type of compensation of magnetic field inhomogeneity, caused by an extraction channel, was proposed by Reich and Löns [2.11]. A computer study of such a structure was made in [2.12].

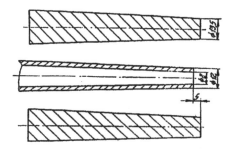

Figure 2.13. Geometry of an extraction channel with two compensation shims above and below it which is used in the $17\,\mathrm{MeV}$ microtron at the Institute for Physical Problems (Moscow).

Figure 2.14 illustrates experimentally measured variation of the magnetic field in the vicinity of the extraction channel, depending on the distance between compensation shims. The pole gap equals in this case 10 cm. In the insert of the figure there is dependence of the relative magnetic field change on the previous orbit on this distance. One can see that it is possible to obtain practically complete compensation of the disturbance induced by the iron extraction tube in the accelerating plane.

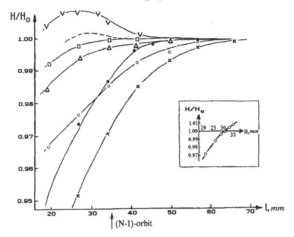

Figure 2.14. Relative magnetic field change along the common diameter of the orbits at different separation of compensation rods from the centre of the extraction tube, pole gap equals 92 mm: • — without shims, ×, o, △, □, ∨, correspond to a distance of the compensation rods from extracting tube 21.5, 26.5, 29, 31.5, and 34 mm; dashed line corresponds to a optimal distance between rods 32 mm. *Insert*: relative magnetic field H/H_0 at the previous orbit position in dependence of a rod distance a from extracting tube.

Two methods could be considered to extract a beam from various orbits. The first one is to move the resonator to shift an orbit of necessary energy to the extraction channel. An advantage of this way is the relative simplicity of the drive, in addition the magnetic system of the microtron need not be changed. Certainly, one can use special pieces of a waveguide of constant length to change the electron energy by a definite number of orbits.

An electron-beam extraction scheme with a moving accelerating cavity was realised by Gridnev *et al.* [2.13] and is shown in Fig. 2.15. Mounted in the microtron chamber are an extraction shunt and a resonator, connected to a waveguide insert with spring contacts, which is introduced into a fixed waveguide channel. The mechanism for moving the accelerating resonator along the entire orbit diameter can be located outside or inside the vacuum chamber of the microtron.

The energy of the extracted electron beam is varied as follows.

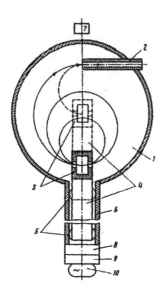

Figure 2.15. Electron-beam extraction. 1) Microtron chamber; 2) extraction shunt; 3) resonator; 4) waveguide insert; 5) spring contacts; 6) fixed waveguide channel; 7) mechanism for moving resonator; 8) phase shifter; 9) ferrite gate; 10) high-frequency generator (magnetron).

The resonator, with the waveguide insert and the attached spring contacts, is installed at distance L from the extraction shunt by means of the mechanism for moving the accelerating resonator along the entire electron-orbit diameter. If distance L is a multiple of the distance l between adjacent orbits ($L = kl$, where $k = 2, 3, ...$), accelerated electrons are extracted from the $(N-1)$th, $(N-2)$th, etc., orbits, and the energy of the extracted beam will vary by $k\Delta E$, where ΔE is the increment in electron energy for one pass through the accelerating resonator. The accelerating resonator and the high-frequency generator are matched by a phase shifter and a ferrite gate.

The waveguide insert, which has a length of 325 mm and an outside cross section of 40×68 mm, is made from a standard copper waveguide with a cross section of 44×72 mm. Bronze spring contacts 20 mm long (the contact tabs are 0.2 mm thick and 4 mm wide) are soldered to one side of the insert over its entire perimeter, and a brass rim for attaching of the resonator is soldered to the other side. Each spring-contact tab creates a force of $\simeq 0.1$ kg/mm2 on the inner wall of the fixed waveguide. The resonator is attached by bolts through lead gaskets to the rim and along with the waveguide insert is connected to the mechanism for moving the resonator along the entire orbit diameter. The accuracy of the resonator setting relative to the median plane of the microtron electromagnet is 0.1-$0.3°$ in angle and 1-2 mm in height.

The movement mechanism consists of two fixed guides, which are attached to the lower pole of the electromagnet, and a movable plate, which moves along the pole between the guides (dove-tail joint). A movable shaft is rigidly connected to the plate (on which the waveguide inset with resonator is located) and is connected through a vacuum seal in the chamber to a reducing gear and a reversible motor. Graphite lubricant is used to reduce friction in the resonator movement mechanism. The parts of the resonator movement mechanism that are located in the microtron chamber are made of non-magnetic materials. The rate of movement of the resonator with waveguide insert relative to the electromagnet pole is 7 mm/min.

The variation in the electrical length of the microwave circuit, when the resonator with waveguide insert is moved, is compensated for by a 10-cm-band bridge phase shifter with a remote drive. The microwave energy is transmitted through a waveguide circuit with a cross section of 44×72 mm and a waveguide insert with a cross section of 40×68 mm. Energy losses are small, since the condition $(\lambda_{cr})_{H_{10}} < 2a$ (where $\lambda_{cr})_{H_{10}}$ is the H_{10}-mode critical wavelength and a is the larger dimension of the inside waveguide cross section) is satisfied and because the spring contacts ensure satisfactory electrical contact between the moving and fixed parts of the waveguide circuit.

The energy of the extracted electron beam is adjusted by moving the resonator along the entire orbit diameter relative to the fixed extraction shunt. Since the adjustment mechanism permits resonator movement without disturbing the vacuum in the microtron chamber or the electrical supply of the resonator, the energy of the extracted electron beam can be changed rather rapidly. The bridge phase shifter is remotely adjusted simultaneously with movement of the resonator. The beam-energy adjustment time between two adjacent orbits is ~ 10 min.

Another possibility is to use movable extraction channel and optics which conserves the position and direction of the extracted beam. A class of such systems was proposed which reduces the necessary movement of the extraction channel to rotation around the point. The basic idea can be described by the following (see Fig. 2.16): with a point-like cathode the trajectory of an equilibrium particle is representing by a set of circles which are tangent in some point P inside rf cavity. So the line PL exists which is tangent to all of the circles. Let the axis of the extraction channel AB be tangent to one circle with center C. Let us choose the point of rotation O of the extraction channel in such a way that AB is mirror symmetrical to PL relatively to OC. Evidently, with such a choice of rotation point the extraction channel will be tangent to any of the circles at a "right" angle of rotation so the beam can be extracted from any orbit. But the position and direction of the extracted beam just at the channel exit are still different for different orbits and, as it is noted above, an optical system must be used to make the extracted beam position and direction become independent of the channel rotation.

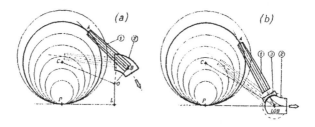

Figure 2.16. Extraction with movable channel: 1 — extraction channel, 2 — bending magnet, 3 — movable part of the magnetic pole.

Let us consider two cases (see Fig. 2.16):

a) with the shoulder OB of non-zero length the necessary magnetic-optical element to get a constant direction and position of the extracted beam is the only bending magnet ensuring the trajectory segment is to be an arc with center O. Of course, the effective edge of the ferromagnetic channel B must be chosen so that OB is perpendicular to AB.

b) with OB of zero length (it means that O lies at the intersection of AB and PL) the bending magnet to conjugate two straight segments of trajectory has a special pole shape, roughly speaking, it is a part of a circle with center O. No conditions apply to the effective length of the extraction channel.

The first type of extraction (*a*) was realized at the microtron in JINR (Dubna) [2.1] and is shown in Fig. 2.17. The extraction tube is rotated relative to the centre of the turning magnet 6. In the changing of the extraction orbit the channel 4 is simultaneously moved along its axis and along the common diameter of orbits to be tangential to the extraction orbit. The construction allows the beam to be extracted from the last 10 orbits, the energy change is from 10 to 25 MeV.

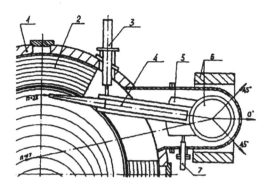

Figure 2.17. Dubna variant of the extraction system: 1 — magnetic core, 2 — coil, 3 — the system of moving the extraction channel to different orbits, 4 — extraction channel, 5 — the system of moving the channel along its axis, 6 — turning magnet, 7 — external driver for the system 5, n — the number of the orbit.

The second approach (b) was used in the microtron of variable energy designed in Novosibirsk [2.7]. To avoid aberrations which would be at the circular pole, the bending magnet pole consists of both motionless and movable parts. The movable one rotates together with the extraction channel, so that the entering beam always 'sees' the straight edge of the magnet. The window in the back magnetic core for moving the extraction channel is rather big. To decrease the azimuthal distortion of the magnetic field there is a movable piece of the back magnetic core. It rotates together with the extraction channel completely filling the window when extracting from the last orbit.

There is also another possibility of electron beam extraction which follows the suggestion of Reich [2.14] and was made in the Swedish 22 MeV microtron for radiation therapy [2.15]. According to this proposal the beam from any orbit can be extracted at the same place in the same direction. By displacing any of the circular orbits, always by the same distance, the common point of the tangent will also be displaced by this distance. Therefore, all electrons from any of the displaced orbits can be extracted at the same point and the beam always leaves the accelerator along the displaced common tangent. In practice this is accomplished by a narrow deflection tube of steel which moves along a straight line, and the centre of the orbit will be displaced by the length of the tube. In this way the final orbit can always be extracted from the magnetic field by a fixed extraction tube.

2.1.3.2　Injection of Electrons from a Microtron to Synchrotrons

Recently more and more high energy electron synchrotrons are being developed in different countries from all over the world. This tendency is governed by the nuclear physics interests in high-intensity and high-quality sources of relativistic electrons, as well as the possibility of using circulating fast electrons for synchrotron radiation production.

The quality of the beam from the preaccelerator is naturally of vital importance for the number of particles that can be accepted into acceleration by the synchrotron. From this point of view the microtron accelerator seems to be very suitable as an injector into a synchrotron. Actually, in many synchrotrons the microtrons are used for supplying an electron beam of 10-50 MeV energy into synchrotrons. There are two machines in Russia — in Troitzk near Moscow [2.16] and in Tomsk [2.17], in the USA microtrons are used at Tantalus [2.2], Surf II [2.18] and Aladdin [2.19], at the University of Lund in Sweden [2.20], in the Centre of Advanced Technology in India [2.21], and at Eindhoven University of Technology the electron storage ring EUTERPE is under construction, which has also a microtron injector [2.22].

Matching a Microtron Emittance with a Synchrotron Admittance

The quality of an accelerator beam is usually illustrated by a phase diagram obtained by plotting the particle transverse momentum or angle of divergence as a function of its distance from the beam axis. These curves usually have an elliptic shape and the enclosed area, which without acceleration is invariant, is called the beam emittance. The two emittance diagrams in Fig. 2.18 represent the microtron beam quality in the radial and axial planes just outside the extraction pipe of the Sweden machine with external injection. These plots are practically the same for all microtrons.

Analogously to the preaccelerator beam emittance, it is possible to define also the synchrotron admittance. In a corresponding phase diagram the limiting values of the lateral positions and associated angles are plotted, at which the electrons can be launched from the inflector or the vacuum chamber walls. These limiting values form a closed curve and the area inside is the admittance.

In order to perform effective capture of electrons from an injector into a synchrotron it is necessary to match the injector's emittance with the synchrotron's admittance. In the ideal case the emittance and admittance have to coincide exactly. In practice it is impossible to obtain their complete coincidence, and the

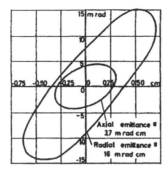

Figure 2.18. Microtron beam emittances [2.20].

problem is to achieve the maximum possible intersection of the phase ellipse of the ingoing beam with the admittance of the main accelerator.

Electron beam injection into the synchrotron during the period of only one revolution is difficult. As a rule, the radius of a synchrotron is around 10 m or more and the time of revolution of relativistic electrons is about $0.1\,\mu s$. Injection by bending the beam directly on to the central orbit would require an electrostatic or magnetic inflection device, which could be pulsed in less than 10 ns. Technically this is extremely difficult and for reasons of intensity a very high peak current is moreover required from the injector during the $0.1\,\mu s$ injection interval.

Therefore, as it was shown by Wernholm [2.20], multiturn injection is thus the only possible method. The beam from the microtron enters the synchrotron vacuum tube through an electrostatic inflector located in a straight section outside the central equilibrium orbit. Injection can start immediately the magnetic field reaches a value at which the equilibrium just clears the inside of the inner inflector electrode. The electrons, since they are injected from a position outside the equilibrium orbit, will perform a betatron oscillation around this latter. In a stationary magnetic field the electrons would sooner or later hit the back of the inflector but in the slowly increasing magnetic field the equilibrium orbit will spiral inwards and some electrons with favourable starting conditions in respect of amplitude and direction will miss the inflector obstruction for a sufficient number of turns to survive.

From the definition of the momentum compaction factor $\alpha = (\Delta R/R_0)/(\Delta p/p_0)$, the rate of spiralling per turn of the equilibrium orbit is found to be

$$\Delta R = \frac{R_0}{B_{0i}} \alpha \frac{dB_{0i}}{dt} \frac{2\pi R_0}{c}. \tag{2.20}$$

Here B_{0i} is the flux density of the magnetic field at the moment of injection. A spiralling rate for the equilibrium orbit is equal to about $0.25\,\text{mm/turn}$.

For effective capture of an injecting beam into the process of acceleration the maximum amplitude of vertical oscillations has not to exceed the half-height h of the vacuum chamber of the accelerator. Thus, the condition of capture is written in the form

$$\sqrt{z_0^2 + \gamma_z^2 R_0^2/n} \le h, \tag{2.21}$$

where z_0 is the amplitude of the vertical oscillations of a particle, γ_s is the angular vertical divergence of the injected beam, R_0 is the radius of an equilibrium orbit, and n is the magnetic field index in the synchrotron. In the plain (z, γ_s) the equation (2.21) is depicted by an ellipse with semiaxes h and $h\sqrt{n}/Ro$ which is a graphical representation of the vertical admittance of the accelerator. For instance, Fig. 2.19 illustrates vertical admittance which is calculated according to formula (2.21) for the $1.5\,\text{GeV}$ Tomsk synchrotron at $R_0\text{=}423\,\text{cm}$, $h\text{=}4.3\,\text{cm}$, and $n = 0.58$.

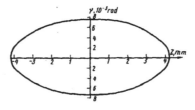

Figure 2.19. Vertical admittance of 1.5 GeV synchrotron in Tomsk [2.23].

While calculation of the vertical admittance is rather simple, the radial admittance of a synchrotron with multiturn injection has a more complicated character, as it depends on the moment of injection. The capture condition can be written in the form

$$\frac{\tau \Delta r}{T} + \Delta r k = \left(\frac{\tau \Delta r}{T} + q \right) \cos \sigma_r k + \alpha \sin \sigma_r k, \tag{2.22}$$

where $\tau = M_i T/\Delta r$ is the time interval from the onset of injection, T is the period of one turn, M_i is the distance along the radius from a momentary orbit to the internal edge of the inflector, Δr

is the orbital radius change in one turn, and k is the number of electron turns in the chamber, and $q = \rho_i - M_i$, ρ_i is the particle radial deflection from a momentary orbit. Besides

$$\alpha = \frac{\sin \sqrt{1 - n}\nu + (\sqrt{1 - n}\, L/R_0) \cos \sqrt{1 - n}}{\sqrt{1 - n}\, \sin \sigma_r}, \qquad (2.23)$$

γ is the radial angle between the velocity of the particle and the momentary orbit at its outlet from the inflector, and the σ_r is the change of the phase of radial oscillations in one element of the periodic magnetic system. The value of σ is determined from the expression

$$\cos \sigma_r = \cos \sqrt{1 - n}\,\nu - (\sqrt{1 - n}\, L/2R_0) \sin \sqrt{1 - n}\,\nu, \qquad 2.24$$

ν is angular spread of the magnetic sector, and L is the length of a rectilinear gap of the magnet.

Let injection begin at $t = 0$ when the momentary orbit is passing through the internal edge of the inflector. At a given time interval from $t = 0$ to $t_i = M_i T/\Delta r$ (time interval of an injection) at the phase plane (q, γ_r) we obtain the condition (2.23) which displays a closed region of capture (synchrotron admittance), restricted by segments of straight lines. These segments depict the edge of an internal electrode of the inflector at different values of k. At τ, which corresponds to the end of injection, the admittance is restricted to right by the internal wall of the chamber or by the edge of the operation region of the magnetic field if this region is smaller than the radial dimensions of the chamber.

Figure 2.20 represents the radial admittances of the Tomsk synchrotron at different times after injection at $\Delta r = 0.3$ cm. As it is well known, the vertical emittance of the microtron is sufficiently less than the horizontal emittance and, as was mentioned above, it is simpler to match the vertical microtron emittance with horizontal admittance of the synchrotron. On the other hand, matching of the horizontal emittance of the microtron with vertical admittance of the synchrotron is also not a problem.

The interval during which an injected beam can survive for many turns starts at the instant when the equilibrium orbit comes just inside the inflector electrode. It ends when the equilibrium orbit reaches a position midway between the inflector and the inner vacuum tube wall, because at the latter position the betatron amplitudes of the injected beam cover the entire radial aperture. Taking the radial spiralling distance to 25 mm gives an injection interval of 10 μs. This interval is usually much longer than the injection pulse, being some microseconds.

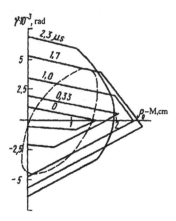

Figure 2.20. Radial admittances of the 1,5 GeV synchrotron at Tomsk at different times after beginning injection at $\Delta r = 0.3$ cm. The broken line corresponds to the phase ellipse of the beam after the inflector.

Thus, the microtron emittance has to be turned by 90° relative to the synchrotron plane. In the Lund synchrotron this was done mechanically by placing the microtron vertically.

Ideally, the injector beam should be 1-2 mm wide, parallel and without energy spread. Its energy should be high, preferably above 5 MeV in order to reduce the influence of remanence in the synchrotron magnets and stray fields in the straight sections. A high injection energy also reduces scattering losses and considerably simplifies the radio frequency acceleration, by reducing the extent of frequency modulation.

The beam, in its transportation from the microtron to the synchrotron, has to pass through some kind of lens system, in which it is given a shape suitable for injection. To put it in phase space terms, the object of the lens system is to transform the microtron beam emittance to fit the synchrotron admittance. As it was mentioned, the emittance area is an invariant, and only its shape can be altered. The microtron is therefore mounted at Lund with the orbit plane vertical, so that the axial beam emittance is coupled to the radial admittance of the synchrotron.

Rotating the Microtron Emittance

Matching the microtron beam emittance with synchrotron admittance can be also done by means of the axially symmetrical magnetic field of a lens, as in Tomsk [2.23].

For simplicity, let us consider the motion of only one electron from a narrow paraxial beam inside a lens, and let its original trajectory forms small angle α relative to the direction of magnetic field lines. In a homogeneous magnetic field an electron would move in a spiral which, projected onto a plane perpendicular to the magnetic field lines is a circle with radius

$$\rho = \frac{cmv_n}{eH_0} = \frac{cmv\sin\alpha}{eH_0},\qquad 2.25$$

where v_n is the component of velocity perpendicular to the magnetic field lines.

In time

$$\tau = \frac{2\pi p}{v_n} = \frac{2\pi cm}{eH_0}\qquad (2.26)$$

an electron will circumscribe one turn of a spiral and the path traversed

$$d = v_t\tau = \frac{2\pi cmv\cos\alpha}{eH_0},\,(2.27)$$

where v_t is the component of velocity parallel to the magnetic field.

In order for an electron to pass a distance which, projected onto a plane perpendicular to the magnetic lines, will be equal to $\pi p/2$ and will turn by $90°$ relative to its initial position, the magnetic field intensity has to be equal to

$$H_0 = \frac{\pi cmv\cos\alpha}{2ed}.\qquad (2.28)$$

In practice the magnetic lens was a one-layer coil which was placed inside the extracting tube of the microtron, and it was 120 mm in length and 16 mm in diameter. The coil was fed from a pulsed supply.

The results of the measurements of the outgoing electron beam emittance, with and without the turning magnetic lens, are shown in Fig. 2.21.

2.2 Cathode

Much effort has been devoted to the development of effective electron emitters during the past forty years. The available space for the electron injector in a microtron is rather limited. From a surface of, say, $0.1\,\text{cm}^2$ a beam in the ampere range has to be created. Commonly, in an experimental set up, the operating vacuum is comparatively poor, and, preferably, the cathode

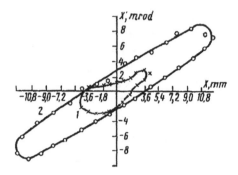

Figure 2.21. Horizontal emittance of the microtron beam under the action of the field of a turning coil (1) and without the coil (2). The area which is restricted by the curves divided by 2π is equal to the emittance of the beam.

should tolerate frequent exposures to atmosphere. Borides of the rare-earth metals, and especially lanthanum hexaboride, meet these requirements and have therefore been used as cathode materials in practically all laboratory microtrons as well as in commercial machines. The scope of this section is to present different aspects of the practical use of LaB_6 in microtron cathodes. Wide consideration of these problems was earlier performed by Wolski [2.24] and by Rosander and Wernholm [2.25].

2.2.1 Properties of Lanthanum Hexaboride

At temperature $T = 1600°C$ and zero field at the cathode, the polycrystalline and [100]-face single-crystal LaB_6 cathodes emit ~8 and 20-22 A/cm^2, respectively. In the absence of ion bombardment, the lifetime of the LaB_6 cathode is determined mainly by three factors: the evaporation of the cathode material, which results in changes of the emitter geometry and, as a consequence, of the optical properties of the system; the destruction of the emitter holder due to boron diffusion; and the operating period of the heater.

Setting the cathode lifetime as the evaporation of a 0.1 mm layer of LaB_6, the period for polycrystalline LaB_6 is ~ 1000 ÷ 1500 h at $T = 1600°C$. Actually, during the evaporation process, different sides of crystallites evaporate differently, and the polished cathode surface becomes rougher. The single crystal evaporates more uniformly, and the cathode surface remains smooth and glossy even after the evaporation of a 0.3-nm layer (in the absence of intense ion bombardment). As the temperature of the LaB_6 single crystal is about 70-100°C lower than that of the polycrystalline emitter at the same current density, the lifetime of the single-crystal emitter is 5-10 times longer.

While chemically inert at room temperature, the lanthanum boride is quite active when hot. The boron atoms easily migrate into most base materials, leaving the metal atoms free to evaporate. The cathodes therefore must be isolated from their supports, and this strongly restricts the freedom of design. However, to a varying degree, a few materials do isolate the boride. At lower temperatures, i.e. 1300-1400°C, rhenium can act as a base material in contact with the cathode. With fillers like $MoSi_2$ and $TaSi_2$ the boride pellet may be brazed to, and reasonably isolated from, the base up to, say, 1500°C. Nevertheless, some deterioration of the cathode takes place. Practically the only material giving satisfactory isolation, also at elevated temperatures, is carbon. If the application so permits, all other materials should be prevented from making direct contact with the boride. The boride-graphite complex can either be held mechanically or brazed to its holder using fillers of e.g. a Ta-Ti alloy.

The lifetime of the cathode is not only determined by evaporation of the material. The hexaboride reacts with the oxygen of the residual gas forming oxides, both from the boron and from lanthanum, which vaporize at about 1600 K. In fact, the cathode temperature is higher and the oxides leave the cathode surface and deposit on the walls of the cavity resonator and, since they are insulators, can charge up and cause arcing. The effect determines the lifetime of the cavity because arcing is irreversible and inhibits cavity operation, and occurs, at 10^{-6} Torr, every 50 h; but at 10^{-7} Torr, this effect occurs only after more than 250 h of operation.

The cathode lifetime can be increased by proper treatment, as it was demonstrated in [2.3]. The directly heated cathode of pressosyntered LaB_6 was a cylinder 3 mm diameter and it was pill housed in a suitable rhenium cylinder with two tantalum legs spot welded to it. The cathode is heated by passing a current through the legs higher than that used during operation in the microtron, in order to reach a temperature of about 2200°C. At this temperature an eutectic alloy is formed between the LaB_6 and the rhenium and the two materials weld together. During more than 500 h of operation in the microtron cavity (vacuum at 10^{-7} Torr) and over 2000 hours at 1800°C in a test chamber (vacuum at 10^{-8} Torr), no sign of poisoning was found on the surface of the emitter.

2.2.2 Energy Relations

Thermal losses of a cathode are due to thermoconductivity by its holders and due to radiation. As a rule, these two ways of energy loss are comparable. The dependence of the density of emission current on temperature is described by Richardson's expression

$$j(T) = AT^2 \exp\left(-\frac{\varphi - e\sqrt{eE}}{T}\right),\qquad(2.29)$$

where φ is the work function of a material used. Note, that the Schottky effect is also taken into account. This dependence can be approximated in a narrow interval of working temperatures $\Delta T \ll T_0$ by a power-type dependence

$$j(T) \simeq BT^n.\qquad(2.30)$$

To find the power index n, let us take a logarithm of eq. (2.30)

$$\ln j = \ln A + 2\ln T - (\varphi - e\sqrt{eE}\exp(-\ln T).\qquad(2.31)$$

Taking into account the condition $\Delta T \ll T_0$ we obtain

$$\frac{d\ln j}{d\ln T} = n \simeq 2 + \frac{\varphi - e\sqrt{eE}}{T_0}.\qquad(2.31)$$

Let us estimate the magnitude of the member which corresponds to the Schottky effect. In the first accelerating mode an electric field intensity for $\Omega = 1.1$ the member $e\sqrt{eE} \simeq 0.2$, whereas $\varphi \simeq 2.6 \div 3.0\,\text{eV}$, and therefore we can neglect the Schottky effect in the temperature interval $1500 \div 1800\,\text{K}$. As it follows from experimental data, in this interval the function $j(T)$ is described with rather good accuracy by power-type expression (2.30) with a power factor $n \simeq 17 \div 20$.

Consider now the dependence $j(P)$ of emission current on heating power. This is in fact the dependence of emission current on temperature because there is functional dependence $T(P)$, i.e.

$$\frac{dj}{dP} = \frac{dj}{dT}\frac{dT}{dP}.\qquad(2.33)$$

If loss of heat is only due to thermoconductivity through the cathode fastenings, then we obtain from the energy balance that

$$\frac{dP}{dT} = \chi S_\chi,\qquad(2.34)$$

where S_χ is effective cross section of the holders and χ is the coefficient of thermoconductivity.

If thermal loss is determined only by radiation

$$\frac{dP}{dT} = 4\sigma S_\sigma T^3 , \qquad (2.35)$$

where σ is the Steffan-Boltzmann constant and S_σ is the area of the heated cathode.

Appreciate an emission current dependence on heating power by power-type function

$$j(P) = CP^k . \qquad (2.36)$$

It follows from (2.34)-(2.35) that in the first case, when thermal loss is due to thermoconductivity,

$$n = k , \qquad (2.37)$$

and in the second case, when energy loss is determined by radiation, we have

$$n = 4k . \qquad (2.38)$$

Certainly, if both ways of thermal loss are comparable, the power factor k lies in the interval $n/4 < k < n$. For lanthanum hexaboride cathode $5 < k < 20$ for temperature interval 1500-1800 K.

Figure 2.22 illustrates the experimentally measured dependence $j(P)$ for two LaB$_6$ indirectly heated cathodes of different shapes. The experimental curves are described well by the power-type law, and the power factor is in the estimated region.

The finite dimension of a cathode surface leads to undesirable bombardment of it by non-resonance electrons. Thus, the electron beam power reentering a microtron cathode increases its temperature, and one edge of the cathode surface is heated most because electrons from the entire surface are bombarding it and some of these electrons, which are emitted from the opposite edge of the cathode surface, have the highest energy.

Of course, there is a way to reduce the dissipating power by using cathodes which are large vertically and small horizontally. However, accelerated electrons could be lost from the beam by such a cathode because of the large amplitude of vertical oscillations. So one must use a special cathode design and cavity transit holes of proper shape and size.

This problem has recently been considered in [2.26]. Below we shall qualitatively estimate the cathode heating for typical microtron parameters: the magnetron pulse power of 2 MW in 3 μs pulses, with a accelerated beam power ~ 0.5 MW in pulses

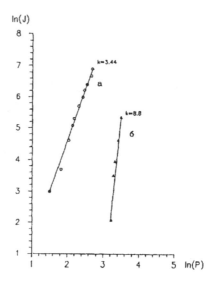

Figure 2.22. Emission current from LaB$_6$ cathode as a function of heating power (accelerating regime with $\Omega = 1.1$) : a — sintered cathode with Ta holder 10 mm length and 0.5 mm^2 cross section; b — single crystal on Ta holder 5 mm length and 2 mm^2 cross section (data of V. Semenov).

and the pulsed current emission $I \simeq 1$ A. A quadratic cathode of a side $a = 3$ mm is at $1600°$C. Take also into account that for LaB$_6$ density $\rho = 4.7$ g/cm^3, specific heat capacity $C = 1.8$ SJ/(g·deg), and thermal heat diffusivity $D = 0.12$ cm^2/s.

Consider a left cathode end where the heating effect is the greatest. The average energy of heating electrons is easy to estimate from geometrical considerations. The maximum radius of electron motion in its half-turn inside a cavity is ~ 20 mm whereas a cathode width is 3 mm. Therefore the average energy of electrons which bombard cathode surface is $E \simeq 30$ keV. Electrons of such energy have a penetration path of $\sim 10 \, \mu$m. On the other hand, the distance of thermal diffusion L during a pulse duration $\tau = 3 \, \mu$s is

$$L = \sqrt{D\tau} \simeq 6 \, \mu\text{m} \, . \tag{2.39}$$

This means that in fact heat is not spread over the whole cathode volume during the rf pulse, i.e. heating is local, and we can consider that the volume of the left end of the cathode heated is $V \simeq 3 \, \text{mm} \times 10 \, \mu\text{m} \times 10 \, \mu\text{m} = 3 \times 10^{-7} \, \text{cm}^3$. Reentering the cathode portion of emitted electrons is some $\alpha \sim 0.1$ of the total emission (reentering of electrons occurs only in the region of rf phase near $\varphi \sim \pi/2$). A left surface strip of a width $l = 10 \, \mu$m is bombarded by electrons which is l/a portion of the total current I (due to the Schottky effect the amplitude of emission current is, roughly speaking, constant). Therefore, an increase of temperature of the left part of the cathode during the microwave pulse duration $\tau = 3\,\mu$ is

$$\Delta T = \frac{EI\alpha\tau(l/a)}{\rho VC} \sim 10\,\text{K}. \tag{2.40}$$

Thus, the increase of cathode temperature on average is some degrees and in conventional microtrons does not play an essential role. Experiments show that in fact electron thermal emission slightly increases during the rf pulse. Nevertheless, if one needs long pulses of accelerated current, the current would increase more and this effect would be bothersome.

2.2.3 Construction of Cathodes

Cathodes for circular microtrons can be divided into two types — directly heated or indirectly heated. Figure 2.23 illustrates the construction of an indirectly heated cathode.

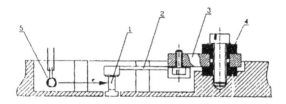

Figure 2.23. Schematic drawing of an indirectly heated cathode used in circular microtrons: 1 — LaB$_6$ cathode, 2 — Ta holder, 3 — current electrode, 4 — ceramic isolator, 5 — W filament.

At different laboratories cathodes, being principally similar, have different arrangements due to the different tasks to be solved by accelerators. Below some examples are considered to illustrate the current situation in cathode design.

Electron or γ-irradiation is the most convenient means of carrying out research on the radiation physics of semiconductors, since the simplest defects (i.e., vacancies and interstitial atoms) are primarily produced by interaction. In order to carry out such investigations, a microtron was constructed and put into operation at the Institute of Semiconductor Physics, Siberian Branch, Academy of Sciences of the USSR, in 1966 [2.27].

Eleven orbits are used in the microtron. The energy of the beam of extracted electrons is controlled smoothly in the range from 2 to 8 MeV by means of two magnetic channels which extract a beam with 4, 5, 6, 10, and 11 orbits through a window into the atmosphere.

The construction of the electron source used is shown in Fig. 2.24. The shape of the grooves on the insulators eliminates the formation of a continuous film as a result of sputtering of the cathode material. The cathode has the dimensions $10 \times 4 \times 2\,mm$. Such a prolonged cathode is used to enlarge the energy region of possible acceleration of elections. The indirectly heated helix is made of tungsten wire with a diameter of $0.2\,mm$ and a length of approximately 9 mm; it is attached to tungsten output leads by spot welds. The cathode holder is made from sheet Ta $0.5\,mm$ thick and consists of two parts which ensure the adjustment of the cathode position relative to the cavity slit. The cathode is fixed by means of a tantalum holder constructed in the form of a spring bracket. The small contact area between the holder and the cathode eliminates the possibility of contaminating the working surface of the cathode, which usually takes place due to the migration of spurious impurities from the holder material.

The electron source is supplied from stabilized dc power supplies. The working temperature of the cathode is controlled and maintained in the 1400 to 1600°C range. In order to stabilize the cathode emission current, dc stabilization of the heating of the tungsten helix with an accuracy of $\pm 0.1\%$ is provided, as well as stabilization of the voltage between the helix and the cathode with an accuracy of $\pm 0.05\%$. The cathode emission current in the indicated temperature range may reach 2.5 A. The variation of the beam current during the process of operation does not exceed 1.2 to 2%.

By controlling the focusing properties of the system consisting of the radial cavity slits and the cathode by applying a small (approximately 10 V) blocking voltage to the cathode relative to the cavity case, one may achieve a reduction of the overall cathode emission current by almost a factor of two without reducing the beam current. The efficiency with which the current in the indirectly heated helix is used amounts to 50 to 60%.

It should be noted that even the very use of an extended cathode leads to an increase in the service life of the electron source, since the density of the emission current is reduced, although under these conditions the capture coefficient is degraded somewhat. The service life of the electron source has been brought up to ~ 1000 h of continuous operation when the indirect-heating helixes are replaced every 80 to 100 h.

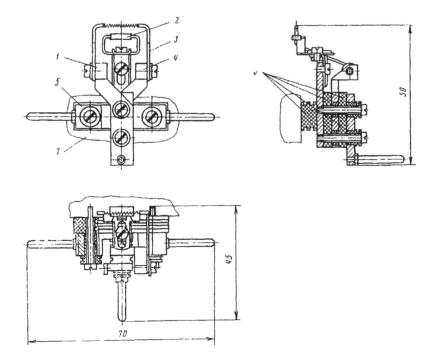

Figure 2.24. Electron source in two projections and cross sections along
the central electrode: 1,4 — current electrodes for heating the helix, 2 —
lanthanum hexaboride cathode which is held by a Ta spring bracket, 3 —
tungsten output leads from the helix, 5 — output leads from the cathode,
6 — insulators, 7 — portion of the cavity case. The parts which ensure
adjustment of the cathode relative to the radial slit of the cavity are visible
in the cross section.

Figure 2.25 presents the cathode for a 10 cm microtron which
is based on modern technology and in which experience of designing electron guns for powerful low-energy electron accelerator
is used [2.28]. In this cathode the LaB$_6$ single crystal [100] face
is used as an emitter and a cylindrical filament is used as the
cathode heater. The latter feature high stability and efficiency
and long service life.

The emitter, made of LaB$_6$ single crystal 1, is fixed in the
graphite holder 2 with an outside diameter of 4 mm. The use
of graphite is known to reduce significantly the boron diffusion
and, as a consequence, to prolong the emitter operating life. The
cathode sleeve 3 is welded with a precisely fitted mount to the
carrying base 6, whose width and thickness are 7 and 0.3 mm,
respectively. The cylindrical filament 4 is made of tungsten wire

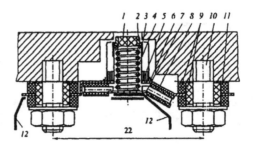

Figure 2.25. Outline of the gun.

0.5 mm in diameter and consists of 8.5 turns with a 0.75-mm step. One lead of the filament is fixed in the cathode sleeve 3 by the tight fit of the enlarged-diameter turn; the other lead is attached to the Ta plate 7, which is insulated from the base with ceramic insulators 8. The cathode sleeve is surrounded by 7 or 8 heat shields 5 providing considerable reduction of heat losses through the side walls of the heater chamber.

The base is insulated from the resonator wall by a set of ceramic tubes and spacers 9, which are mounted on two studs 10 inserted into the resonator wall. Owing to the exact installation of the studs and precise tooling of the gun parts, planar alignment of the cathode is not required. The depth to which the cathode is embedded can be adjusted with the choice of thickness of spacers 11. To carry the current to the cathode heater, Ta strips 12 welded to the heater-fixing plate and the base are used.

In the left- and right-hand halves of Fig. 2,25, two different methods of attaching the gun to the resonator are shown; the attachment shown in the right-hand half was found to be the best.

The results obtained at the testing unit allow us to predict that, in the limited space-charge mode at a temperature of 1600°C and acceptable emitter evaporation of 0.1 mm, the cathode will have an operating life of $\sim 1500-2000$ h. Under these conditions, the cathode provides a current of ~ 1 A.

Since the filament has a cylindrical spiral shape, the magnetic field produced by the filament can affect the motion of the electrons emerging from the cathode. The calculated maximum magnetic field and its components along the surface of the cathode described are presented in Fig. 2.26. The conventional technique for synchronizing the pulses of a high-frequency field when the ac heater current crosses zero permits the elimination of trajectory distortions.

This cathode was successfully tested in the microtron, where

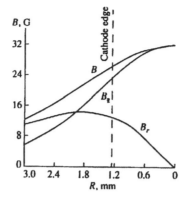

Figure 2.26. Magnetic field along the cathode surface calculated at the filament current amplitude of 20 A. B, B_z and B_r are the total field and the normal and tangential field components, respectively.

it worked during $> 200\,$h without a change in its characteristics.

Above we considered constructions of cathodes located on the resonator cavity in the first and the second acceleration modes. It is also possible to inject electrons from external guns in which LaB$_6$ is also used. Electron guns used in external injection are widely developed in Sweden.

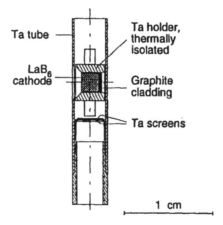

Figure 2.27. Circular microtron cathode arrangement (from [2.25]).

Figure 2.27 shows a 2mm diameter cathode heated by electron bombardment from an auxiliary electron gun (not shown) above the cathode. The LaB$_6$ pellet in its graphite cap is brazed in the small Ta piece with a 65/35 Ta-Ti filler, and, to get low heat contact with the surrounding 4 mm Ta tube, the whole assembly is mounted with the aid of Ta strips. About 30 W of heating power is needed to raise the temperature to about 1650°C, necessary for 1÷3 A emission current. The life of the cathode is in the order of 500 h. Also not shown in the figure is the anode of the gun, a 12/11 mm diameter Ta tube surrounding the cathode tube to form a coaxial system.

A rhenium ribbon, 2 × 0.15 mm, is used as the heater of the gun indicated in Fig. 2.28. For a long time, this type of gun served as the injector of the racetrack microtron at the KTH. The cathode is a 2 mm LaB_6 pellet, as usual cladded with graphite. It is brazed with Ta-Ti into a shallow Ta cup (not shown) spot-welded to the Re heater. The gun, normally operated with a few hundred mA emission, works well, but its life is short. To avoid slackening, the ribbon is stretched by a moderate force (about 1 N) applied to the heater lead, and, due to the limited tensile strength of rhenium at the operating temperature, the heater ribbon is pulled in two after about 150 h of operation.

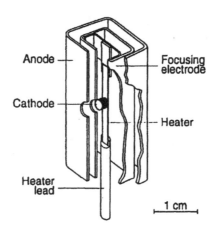

Figure 2.28. Rhenium ribbon heated racetrack microtron gun arrangement (from [2.25]).

2.2.4 Photoemission Studies of LaB_6

During the last decade a lot of experiments were performed to investigate the properties of photocathodes for their use as an electron source for rf accelerators. Because the electrons are emitted via the photoelectric effect, the source is not subject to the thermal emittance limits inherent in thermionic cathodes.

These photodiodes can potentially produce much brighter election beams than are now available. Applications that would benefit from high-brightness electron beams include wakefield accelerators, free-electron lasers, and efficient microwave generators.

It is now clearly understood that laser-illuminated photocathodes are capable of producing electron beams with a current density of more than $200 \, A/cm^2$.

In the simplest mode of operation, in which the applied voltage is dc, a laser is pulsed repetitively onto a photoemissive cathode. The current pulse follows the laser intensity, and, in general, is not space-charge-limited. The electron pulse is then accelerated out of the device by a constant voltage.

LaB_6 as well as metals have proved to be robust and long-lived materials for photocathodes. Their primary drawback is their relatively low quantum efficiency, which necessitates the use of a more powerful driving laser with short wavelength to produce a high photocurrent. The goals for high quantum efficiency and long lifetime are somewhat contradictory.

Photoemission studies on LaB_6, using a nanosecond KrF excimer laser, were described in [2.29]. The laser used in the experiments, a 10 Hz KrF excimer laser, supplies 25 ns FWHM pulses at 248 nm (photon energy 5.0 eV). The maximum pulse energy is 750 mJ. The laser beam is directed through an optical system onto the cathode at an angle of approximately 70° to the normal of the surface. The photoelectrons emitted from the cathode are accelerated across the cathode-anode gap, and collected by the Faraday cup as the photocurrent.

It is now well known that quantum efficiency is sensitive to the surface quality of the photocathode and it is necessary to process the photocathode surface before the experiment [2.30]. The surface of the photocathode is lapped with a diamond compound, washed in solvents and ultrasonically cleaned before installing into the cell. After installation, surface heat treatment (10^{-8} Torr, 1400° C) is used for LaB_6.

Figure 2.29 shows photocharge versus irradiated energy. Using a least squares fit the photocharge is linear with energy. The experimental results show the efficiency of LaB_6 can be enhanced about an order of magnitude by surface heat treatment. The parameters of the treatment are: vacuum better than 10^{-8} Torr, temperature about 1400° C and heating time more than 20 min. The key is that the vacuum must be 10^{-10} Torr during treatment. It has been found that an efficiency of 6.1×10^{-4} is achieved after treatment, but it will decrease with a decreasing cathode temperature and stabilize at about 2×10^{-4} after 4 hours. The reason is that the surface will absorb residual oxygen in the cell while cooling [2.31].

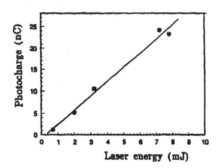

Figure 2.29. Photocharge emitted from LaB$_6$ versus laser pulse energy (from [2.29]).

2.3 Injection of Electrons From an External Source

2.3.1 Injection From an External Gun

Injection of electrons into an accelerating cavity from an external gun was first suggested and performed in Sweden [1.6,1.7]. Electrons were introduced into a toroidal cavity by a pulsed high-voltage coaxial gun. By placing the electron gun outside the cavity, it is possible to adjust the emission voltage between the cathode and the anode independently of the phase of the accelerating rf field. This feature allows for a beam current higher than is possible in microtrons using a hot cathode inside the cavity, because there is no need for rf power to accelerate non-resonance electrons inside the cavity.

However, using a toroidal cavity as an accelerating element does not allow to change the energy of accelerated particles smoothly because the magnetic field has to be equal to the cyclotron field. Recently, a combination of the second type of acceleration in a rectangular cavity with an external electron gun was developed in Japan [1.8].

To achieve resonance, the conditions of synchronous motion (1.3) must be rewritten for external injection in the form

$$T_1 = \frac{2\pi(mc^2 + eV_g + E_i + E_r)}{eBc^2} = \frac{\mu}{f}, \qquad (2.41)$$

$$T_N - T_{N-1} = \frac{2\pi E_r}{eBc^2} = \frac{\nu}{f}, \tag{2.42}$$

where E_i is the energy that electrons gain inside the accelerating cavity before entering the first orbit, E_r is the energy gain, and V_g is the voltage applied between the cathode and the anode. These equations give

$$E_r = \frac{\nu}{\mu - \nu}(mc^2 + eV_g + E_i), \quad B = \frac{2\pi f E_r}{ec^2\nu}. \tag{2.43}$$

The kinetic energy of the Nth orbit is

$$E_N = E_r \left(N + \frac{\mu - \nu}{\nu} \right) - mc^2. \tag{2.44}$$

By using external injection the injection energy $(mc^+ eV_g + E_i)$ can be optimized because the voltage V_g can be adjusted.

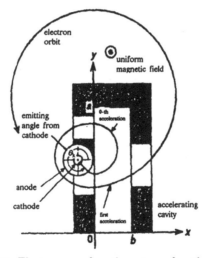

Figure 2.30. Electron gun location at accelerating cavity.

The electron gun consists of a cathode and an anode. The cathode is a cylindrical lanthanum hexaboride button and is fixed in position by a cylindrical tantalum holder. The anode is a tantalum cylinder with a slit. The holder and the anode are formed into coaxial cylinders and are placed outside the cavity as shown in Fig. 2.30. The holder can be rotated on its axis by a motor so that the injection angle of the electrons can be adjusted under operation. This kind of electron gun has been used in Swedish microtrons. The dimensions and positions of the electron gun were determined by means of computer simulation.

The circuit of the electron gun is shown in Fig. 2.31. The cathode is heated by the bombardment of electrons emitted from a tungsten spiral filament. Electrons are emitted from the surface of the cathode when a pulse of high-voltage V_g (pulse duration: $4\,\mu s$, pulse repetition frequency: $50 \div 250\,Hz$) is applied. The emission current can be varied by changing the electron bombardment power and is usually from 0.5 to 2 A. Moreover, the voltage V_g can be adjusted within the range of $10 \div 30\,kV$ to optimize the injection energy of electrons.

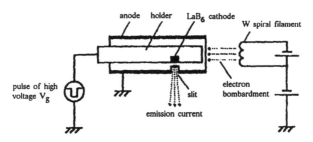

Figure 2.31. Circuit of the electron gun.

The cavity studied has three apertures in its walls. Electrons emitted from the gun first enter the cavity through the entrance aperture, then they are accelerated and exit from the cavity through the supplementary aperture. This acceleration was called the zeroth acceleration by the authors. Then, the electrons run into the cavity again through the entrance aperture and exit from the cavity through the exit aperture on the opposite wall, which is called the first acceleration, and, afterwards, this acceleration is repeated (the nth acceleration, $n > 2$).

Numerical analysis of electron orbits in this type of injection was also performed [2.32]. The calculation region was separated into three parts as follows: 1) inside the electron gun; 2) inside the accelerating cavity; and 3) outside the cavity and the gun.

The calculation of the electron trajectories starts at the cathode, with the emitting angle from the cathode θ_k as an initial condition. Another initial condition is the phase of the rf field at the time of election injection into the cavity under the zeroth acceleration, i.e. the initial injecting phase $\varphi_{in}(0)$. The electron orbits were calculated for each combination of initial conditions $\theta_k \varphi_{in}(0)$ until the desired nth acceleration was accomplished or the electrons ran into the wall of cavity.

The region for stable acceleration under the initial conditions $(\theta_k, \varphi_{in}(0))$ is shown in Fig. 2.32. Parameters other than

$(\theta_k, \varphi_{\text{in}}(0))$ were optimized so as to maximize the region in the $(\theta_k, \varphi_{\text{in}}(0))$ plane corresponding to stable acceleration until the 20th orbit.

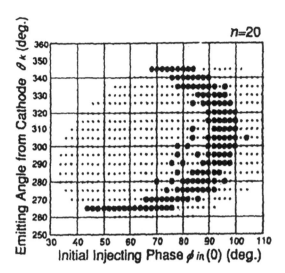

• : stable up to 20th orbit

· : energy lower than $0.95 \times U_{20}$ or collision with the cavity wall in 2nd-19th orbits

Figure 2.32. Stable region under the initial conditions of the microtron. The calculation was performed for each combination of $(\theta_k, \varphi_{\text{in}}(0))$ represented here, every $5°$ in θ_k and $2°$ in $\varphi_{\text{in}}(0)$.

From Fig. 2.32, it is obvious that there is a definite region of stable acceleration until the 20th orbit. Accordingly, stable acceleration appears to be possible in the type of microtron we investigated, based on this simulation. In Fig. 2.32 the stable region for the combination of $(\theta_k, \varphi_{\text{in}}(0))$ for the 20th orbit forms a crescent-like shape with a center axis at about $\theta_k = 305°$. In addition, the distribution of all electrons that were accelerated until $n \geq 2$ has an ellipse-like shape, and its range of θ_k is about the same as that of the crescent shape, which is about $80°$.

Figure 2.33 shows the dependence of the total width of the stable range of θ_k on $\varphi_{\text{in}}(0)$, which is derived from Fig. 2.32. A peak appears around $90° \leq \varphi_{\text{in}}(0) \leq 92°$, which we will discuss later.

Figure 2.34 shows the dependence of the initial energy ratio

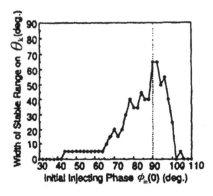

Figure 2.33. Dependence of the width of the stable range of θ_k on $\varphi_{in}(0)$.

$U(O)/U_r$ on the initial conditions $(\theta_k, \varphi_{in}(0))$, where the initial energy $U(0)$ is the total energy of the electron before the first acceleration. In Fig. 2.34a the dependence of $U(0)/U_r$ on the injecting phase $\varphi_{in}(0)$ is shown with θ_k as a parameter. Similarly, the dependence of $U(0)/U_r$ on the emitting angle θ_k from the cathode is shown in Fig. 2.34b with $\varphi(0)$ as a parameter.

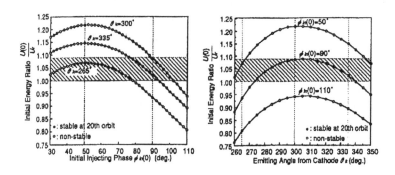

Figure 2.34. Dependence of initial energy of electron on $(\theta_k, \varphi_{in}(0))$: *left* — dependence on $\varphi_{in}(0)$, and *right* — dependence on θ_k.

From these figures, we can see that stable acceleration is most likely to occur when $1.0 < U(0)/U_r < 1.09$ (the shaded portion) and $\varphi_{in}(0) \geq 44°$. Thus, the initial energy $U(0)$ is critical for attaining stable acceleration. Since the dependence of $U(0)/U_r$ on θ_k has a peak around $300° \leq \theta_k \leq 310°$, as shown in Fig. 2.34b, the $\varphi_{in}(0)$ range ensuring that $U(0)/U_r$ is in the shaded portion moves to a higher level as θ_k increases up to $300°$, then moves to a lower level as θ_k continues to increase over $310°$. This relationship between the $\varphi_{in}(0)$ range and θ_k agrees with the stable region shown in Fig. 2.32, the center of which lies at about $\theta_k = 305°$. It is also consistent with the crescent-like shape of the stable region in Fig. 2.32.

In addition, Fig. 2.34b shows that the curve for $\varphi_{in}(0) = 90°$ has the largest range lying within the shaded portion, which is again consistent with the $\varphi_{in}(0)$-dependence of the stable range of θ_k shown in Fig. 2.33.

In Fig. 2.35, the relation of the angle and position of the beam injected into the cavity in the zeroth acceleration is shown to clarify the relation between θ_k and $U(0)/U_r$ shown in Fig. 2.34. In Fig. 2.35, the injecting angle θ_{in} defined in the figure decreases as θ_k increases when $\theta_k < 315°$ and increases when $\theta_k > 315°$. This is considered to be a peculiarity of this particular type of microtron. In the case of a cathode placed by itself inside the cavity, the value corresponding to the injecting angle θ_{in} is always around zero. The injecting position y_{in}/a slightly decreases as θ_k increases when $\theta_k < 270°$ and increases monotonously when $\theta_k > 270°$. These characteristics of θ_{in} and y_{in}/a can be predicted from the geometrical consideration.

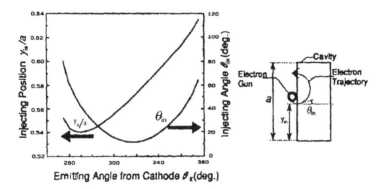

Figure 2.35. Dependence of injecting position and angle into the cavity in the zeroth acceleration on θ_k.

As the injecting angle θ_{in} decreases, the time needed by electrons to pass through the cavity becomes longer, resulting in the increase of $U(0)$. On the other hand, since the electromagnetic field distributed in the cavity has a peak at $y_{in}/a = 0.50$ in the E_{010} mode, the initial energy $U(0)$ decreases as y_{in}/a becomes larger than 0.50. From these results for y_{in}/a and θ_{in}, the dependence of the energy ratio $U(0)/U_r$ on θ_k, shown in Fig. 2.34, can be clearly understood.

Experiments performed in [1.8] show that with electron injection from an external gun in the second accelerating mode and with 2.2 MW microwave power the beam current 80 mA can be achieved at 9.5 MeV.

2.3.2 Electron Injection from a Source Located Outside a Microtron

Certainly, external electron injection can be made from a source located outside a microtron. Such an injection mode is of interest for accelerating polarized electrons. Polarized electron beams are of great interest for different scientific experiments and now such beams exist in many synchrotrons. Mainly, linear accelerators and microtrons are used as injectors for introducing beams into synchrotrons. When polarized electrons are accelerated in linacs the degree of polarization is maintained but the polarization vector rotates in the plane perpendicular to the accelerator axis, causing certain difficulties during the injection of particles into a synchrotron. A microtron can also accelerate a beam with transverse polarization, and in this case the polarization vector does not change direction. As an injector for a synchrotron, the microtron also has the advantage over linacs in the energy range $5 \div 20$ MeV that the electron beam has a small energy spread. In operation with a source of polarized electrons, however, external injection into the microtron is necessary since in order to obtain a beam with good parameters from the source (e.g., current ~ 300 mA at a 45% degree of polarization) an ultrahigh vacuum must be ensured in the source, the emitting surface must be prepared specially, and a system must be provided for rotating the polarization plane, etc.

The microtron with external injection, described in [2.33], was just intended for use as an injector of polarized electrons into a synchrotron and for solid-state research. It was calculated by Kanter [2.34] that there is an accelerating regime in which the injector is taken out of the guide field of the microtron and electrons are transported to the cavity by means of a magnetic channel. The small region of phase capture of the accelerating rf field ($\sim 10\%$) and the large vertical losses due to injection into the retarding field of the cavity make this accelerating regime low in efficiency. Beam injection at an angle to the axis of the toroidal cavity makes possible an increase in the phase region of the electrons captured in the accelerating regime, up to $\sim 30°$. Calculation of the regime for the case of injection into a cylindrical cavity [2.35] shows that a capture region of up to 40° can be obtained. The main fundamental difficulty in realizing a regime of external injection is the small value of the displacement between. the circulating orbits and the injected beam. This problem has been solved theoretically by correcting the first orbit with the aid of solenoid magnetic coils, producing local magnetic fields in the region of the trajectory.

A schematic diagram of a microtron with external injection (MEI) is given in Fig. 2.36. To study the potentialities of the accelerating regime-the source of polarized electrons was simulated with a gun that had a thermionic cathode operating at 150-200 keV and a Pierce optics system.

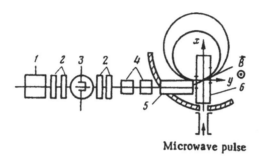

Microwave pulse

Figure 2.36. Schematic of microtron with external injection: 1 — gun, 2 — quadrupole doublet, 3 — measuring unit, 4 — corrector of beam position, 5 — injection device, 6 — accelerating cavity.

An important unit in the operation of the MEI is the system for transporting the beam to the cavity. The following major requirements are placed on the injection unit. Transport of the beam through the region of the uniform and edge magnetic fields should be accomplished so that the instability of the coordinates of beam entry into the cavity does not exceed ±0.3 mm. The minimum perturbation of the magnetic field in the region of the orbits should be 0.2% for a microtron with 10 orbits. The channel wall in the cavity should have a thickness ≤ 1.5 mm, and the beam depolarization in the channel should be minimal.

The design of an injection unit that satisfies these requirements is illustrated in Fig. 2.37 and contains a septum magnet and a system of ferromagnetic channels. The septum magnet is a coaxial line of rectangular cross section and length 750 mm and has a 3 × 4 mm aperture. The length of the magnet is limited by the size of the region of uniform field. The region of pulsed field in the magnet aperture and the region of the orbits of the microtron are separated by a thin wall (septum) of thickness 1.5 mm. The central bus 13, with a 7 × 36 mm cross section, is formed inside the coaxial and is connected to the central bus 7 of the flat coaxial which then passes into a cylindrical vacuum-tight coaxial 6. The coaxial construction in general does not fully eliminate external fields and their magnitude is determined by both the geometry of the system and the frequency of the supply current.

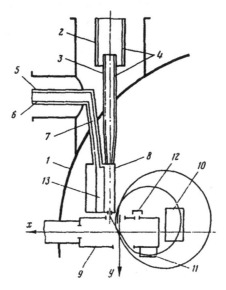

Figure 2.37. Cross section of injection unit in the median plane: 1 — microtron chamber, 2, 3 — ferromagnetic shield-channels, 4 — auxiliary ferromagnetic shield, 5, 6 — elements of vacuum-tight current lead-in; 7 — central bus of flat current lead-in, 8 — septum magnet, 9 — cavity, 10, 11 — correction coils, 12 — Faraday measuring cage.

The ratio of the outer field H_{out} and inner field H_{in} inside the magnet is described, for fairly high frequencies and an outer bus thickness Δ not exceeding the thickness δ of the skin layer in it, by the formula

$$\frac{H_{\text{out}\,m}}{H_{\text{in}\,m}} = \frac{p_m \delta^2}{2\Delta(1 + \frac{2}{3}q_m\Delta)^{1/2}} \exp\left[i\left(\pi - \operatorname{arctg}\frac{2\delta^2}{\Delta^2}\right)\right] \qquad (2.45)$$

where m is the number of the field harmonic, $q_m = \pi(2m+l)/x_0$, and x_0 is the height of the outer bus. From this relation we see that the magnitude of the outer field decreases with increasing height of the central bus and increasing frequency of the supply current. When the ratio of the outer field to inner field is 0.2% the height of the central bus should be 36 mm while the height of the outer bus should be 54 mm. In the operating aperture of the magnet the magnetic field strength is 1130 Oe. Measurements with the aid of an inductance pickup indicate that during the leading edge of the supply-current pulse (15 μs) and the duration of the flat top (6 μs) the outer field near the septum does not exceed the permissible value and is equal to 0.18%.

The mean is guided in the region of highly inhomogeneous field of the microtron by means of two-layer ferromagnetic shields. The outer shield-channels 2, 3 (made of steel) and the inner shield-channels (made of five layers of permalloy) ensure attenuation of the magnetic field on the axis to 0.3 Oe. The wall thickness and length of the channel 2 are 10 and 300 mm, respectively. The ferromagnetic channel 3 with a wall thickness 2 mm and internal diameter 8 mm is joined to the septum magnet through an entrance drift aperture diameter 6 mm.

The value of the pulsed field, which sags in the entrance drift aperture the magnet, depends on its diameter. The dependencies of the magnetic field component B_z the y axis for the drift aperture of the magnet and the end of the channel 3 are of practically the same nature and have opposite signs. This allows the septum magnet and ferromagnetic channel to be joined together without introducing significant perturbations into the trajectory of the low-energy beam. The resulting magnetic field in this region is ∼ 2 Oe.

Analysis of the vertical and radial motion in the microtron determines the principal characteristics of the injected beam: height 3 mm, width 4 mm, angular divergence 0.04 rad, injection energy 130-180 keV, and entrance angle 25-35°. The increase in the horizontal beam size is limited by the possibility of bypassing the top corner of the cavity and by ensuring the septum magnet is bypassed. The angle of entrance into the cavity is changed in this way. Beam shaping at the entrance into the cavity was accomplished with a quadrupole doublet and two pairs of correction coils, making possible parallel translation of the beam. The quadrupole lenses have a length of 30 mm and a maximum gradient of 40 Oe/cm. The choice of quadrupole lenses is based on the minimum depolarization of the beam in comparison with other types of focusing [2.36].

The efficiency of external injection was checked for the following accelerating-regime parameters: width of accelerating gap of cavity 1.815 cm, ratio of electric field strength in accelerating gap of cavity to magnetic field strength of bending magnet 1.00-1.11 (in the Gaussian system of units), and ratio of· fields of bending magnet to cyclotron field $H_0 = 2\pi mc^2/e\lambda$ equal to 1.0-1.3. The duration of the flat part of the injection pulse was $10\,\mu s$ while the duration of the microwave pulse was $4\,\mu s$, which permitted reliable synchronization. With the aid of a Faraday cage 12 set up on the cavity cover opposite the auxiliary window the current injected into the cavity was monitored. In the absence of a microwave field, upon leaving the septum magnet electrons move in the magnetic field of the microtron and impinge on the Faraday cage. Oscillograms of the accelerated and injected beam current are shown in Fig. 2.38.

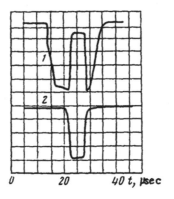

Figure 2.38. Oscillograms of current pulses: 1 — injected current, 2 — accelerated current in ninth orbit. Scale value along the ordinate axis: 1 — $40\,\text{mA}$, 2 — $4\,\text{mA}$.

The main fraction of the losses caused by the small displacement between the circulating orbits and the wall of the septum magnet. All of this leads to deterioration of the capture coefficient, which amounts to $\sim 7\%$. The size of the entrance and exit drift apertures in this case was 0.7×1.7 and $1.2 \times 2\,\text{cm}$, respectively.

The installation of correcting coils permitted the capture coefficient to be raised 10% and a current of $20\,\text{mA}$ to be obtained in the ninth orbit. The coils are shown in the plan view in Fig. 2.37. Each coil is divided into two parts with a $0.8\,\text{cm}$ gap for passage of the beam. The opposite ends of the coils adjoin the magnet poles. The coil 11 has 1000 ampere-turn and the direction of the field in it is opposite to that of the microtron field; the coil 10 has 400 ampere-turns and the field in it coincides with the microtron field. The fields in these coils were $1.25 H_0$ and $1.09 H_0$, respectively. Displacement of the first orbit by $2.5\,\text{mm}$ from the septum magnet along the common diameter of the orbits appreciably decreases the loss of particles.

In order to determine the possibilities of variation of the energies of the accelerated electrons the coefficient of electron capture during acceleration by the magnetic field of the microtron was studied for a fixed injection energy. From the results of the experiment it follows that a maximum capture coefficient of 10%, can be obtained from a microtron field equal to $(1.0\text{-}1.3)H_0$ when the injection energy is varied from 130 to 180 keV. When the vertical size of the injected beam is increased to 5 mm the capture coefficient is decreased to 7% while the angular spread is ±0.04 rad.

The results of these experiments confirm that there is a real possibility of accelerating a polarized beam. Moreover, the application of external injection substantially enhances the reliability of operation of the accelerating cavity. The introduction of regimes with a magnetic field of $2H_0$ and improved transport systems allows the parameters of microtrons to be increased considerably, and this will appreciably expand their range of application in science and engineering.

2.4 Development of Accelerating Regimes

During recent decades, many recalculations of previously found accelerating regimes have been performed, searches of new regions of stable electron acceleration, new experiments on electron acceleration in a microtron have been carried out, and ways for increasing electron capture and focusing have been proposed and investigated. All these problems are considered in this chapter.

2.4.1 Schemes for Correcting Orbits in a Microtron

It is evident that, when in the second type of acceleration the number of the orbit is increased, the angle of inclination of the trajectory of electrons with respect to the resonator axis changes sign. This rotation of the trajectory is caused by the action of the high-frequency magnetic field inside the resonator and the action of the radial electric field near the transit apertures. As a result of the rotation, the electron beam travels flush against the edges of the apertures, and many particles are lost on the resonator covers. Besides, the losses increase if the guiding magnetic field is slightly asymmetrical relative to the vertical plane passing through the common diameter of the orbits. In this case, the orbits drift along their common diameter (i.e., perpendicular to the resonator axis).

This angular drift remains, even in a completely uniform magnetic field. The drift leads to a weakening of the vertical focusing of the particles in connection with the fact that they travel close to the edge of the entry transit aperture where the vertical electric field that focuses the particles is small.

In order to intensify the vertical focusing, Luk'yanenko and Melekhin [2.37] proposed correcting the orbit by means of local current elements situated near the resonator in the region of 'crowding' of the electron orbits (see Fig. 2.39). Near the accelerating resonator, two identical current elements which impart a radial momentum to the accelerated electrons are placed symmetrically on either side of the accelerating resonator. Each element has a coil whose axis is oriented along the guiding magnetic field. The elements are placed directly inside the vacuum chamber and have a narrow cut in the median plane through which the electrons pass. The turns of the coils are wound tightly on the magnet poles, and therefore the reverse magnetic flux created by the correcting element closes its path through the poles and the return magnetic core of the accelerator electromagnet. Because of this, there is only the internal field of the current coils in the working volume, and this field is thoroughly localized; thus, the elements may be placed practically flush against the resonator without perturbing the motion of the electrons inside the resonator on the initial segment of the trajectory. Since the orbits 'crowd' near the resonator, it follows that coils having a small cross section act simultaneously on a large number of orbits.

If the fields created by the coils are identical in magnitude and sign, then their resultant action is equivalent to the action of a local perturbation applied at the spot where the accelerating resonator is situated. Consequently, in the absence of the resonator an angular drift of the orbit would take place. However, if the fields in the coils are identical in magnitude but opposite in sign, then the rotation of the orbit will not change, but the orbits will drift (if there are no other perturbing factors) along the common diameter.

By selecting the shape of cross section of the coil, in the required manner and using two pairs of coils, one may completely eliminate both the angular and radial orbital drifts introduced by the high-frequency field and the nonuniformity of the guiding magnetic field. Using several pairs of coils, one may act on a separate group of orbits with each pair and thereby achieve finer correction. However, experiments have shown that one pair of coils suffices.

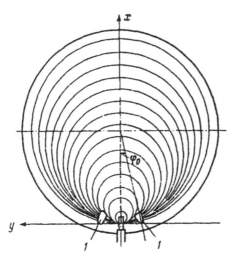

Figure 2.39. Mutual arrangement of the accelerating resonator and the correcting current elements (1).

The correction scheme described not only increases the microtron efficiency but also increases its operating reliability and ensures good replicability of the results. Without correction, the requirements governing fabrication and assembly of the electromagnetic parts are extremely rigorous, since the skewing of the upper pole by several hundredths of a millimeter already causes a nonuniformity that significantly reduces the fraction of particles accelerated to the final energy. Correction of the orbit lowers the indicated requirements substantially.

This method is now used in practically all circular microtrons because they give regulation effect on the orbit position and consequently they, in particular, simplify the process of extraction of electrons.

2.4.2 Motion in a Nonhomogeneous Magnetic Field

As we have seen, inhomogeneity of a guiding magnetic field disturbs radial-phase motion of electrons and can lead to losses of particles. The greater the number of orbits, the higher its influence (see (2.4)). Nevertheless, the close tolerance for magnetic field inhomogeneity relates only to field asymmetry relative to vertical plane incorporating a common diameter of orbits. If the field distribution is symmetrical relative to this direction and the field is two-dimensional, i.e. it is not varied along the common diameter, the inhomogeneity can have higher value. According to the proposal of V. Melekhin and calculations performed [2.38], a magnetic field with such a symmetry, falling towards the magnet edges (roof-like asymmetry), can be used in a microtron for increasing radial focusing.

This idea was tested experimentally by Luk'yanenko and Melekhin [2.39] at the IPP 30MeV microtron with the first type of acceleration. The magnetic field inhomogeneity was performed by means of a set of four steel plates 0.36 mm thick which are mounted inside a vacuum chamber on top and bottom poles. The profile of such shimming plates is shown in Fig. 2.40 by a dashed line. In the same figure the resultant magnetic field distribution is also shown. It can be seen, that the maximum inhomogeneity equals 2%.

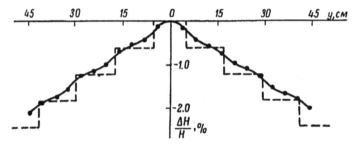

Figure 2.40. Radial distribution of magnetic field with shimming plates that was measured perpendicular to common diameter. The dashed line represents a profile of shimming plates relative to the gap height units.

Authors have shown that in a guiding field with such asymmetry electron acceleration occurs practically with the same parameters (efficiency of electron capture, magnitude of accelerating beam, its dependence on energy) as in a homogeneous field. In the case of slight asymmetry of inhomogeneity relative to the common diameter, radial drift of orbits exists which can be removed by correcting coils.

According to calculations [2.38], the introduced inhomogeneity leads to a drift of equilibrium phase $\sim 4°$, but this drift could not sufficiently change the acceleration process, and this was confirmed experimentally.

The main conclusion of these experiments is that in the decreasing magnetic field the radial motion becomes more stable. Figure 2.41 illustrates this fact, showing the dependence of the magnitude of an accelerated electron beam on field asymmetry relative to the common diameter with and without roof-like asymmetry. It is seen that in a roof-like asymmetric field the field asymmetry decreases electron losses during acceleration.

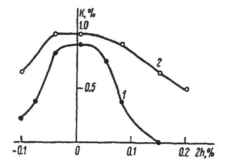

Figure 2.41. Relative dependence of electron beam on left–right field asymmetry in quasi homogeneous (1) and roof-like asymmetric (2) magnetic fields (from [2.39]).

There is another essential feature of electron motion in an inhomogeneous field. In a roof-like field the radial divergence has to be smaller. In an inhomogeneous magnetic field it is approximately four times less than in a homogeneous field. This was qualitatively confirmed experimentally by measuring the radial dimension of a beam.

2.4.3 Second type of acceleration with increased resonator thickness

In the second type of acceleration the principal parameters are as follows: resonator thickness $l = 1.45$, cathode coordinate $x_0 = 0.2$, and ratio of the inductance of the guiding magnetic field to its cyclotron value $\Omega = 1.8$. Acceleration modes in which the range of captured phases reaches 0.8 and the capture coefficient is 25% have also been found in calculation. The share of useful power required for resonance-electron acceleration in the injection region rises to 40%. These modes differ when the resonator thickness increases.

One such mode, with parameters $l = 1.65$ and $x_0 = 0.4$, was tested experimentally [2.40], but the capture coefficient was approximately the same as in the conventional mode, and did not increase as expected. Despite this, acceleration modes with an increased resonator thickness are of practical interest. According to calculations, particles injected in negative phases corresponding to a rising electric field are captured for acceleration in these modes, which increases the efficiency of using of microwave power in the initial acceleration region [2.41]. The similar first type of acceleration by injecting electrons in negative phases successfully operates [2.42]. Moreover, the relative ohmic losses in the resonator are reduced with an increase in l. All of this must result in an increase in the current of the accelerated beam without an increase in the microwave power supplied. Another advantage of the acceleration modes in question is a reduction of the strength of the accelerating electric field for a given energy gain, which corresponds to an increase in the electric strength of the accelerating gap.

More detailed numerical calculations of acceleration modes with an increased resonator thickness (with $l > 1.65$) were performed by Aleshin and Doronin [2.43]. The calculations showed that acceleration of the second type could be obtained with a resonator thickness of up to $l = 1.9$. With a further increase in thickness, the electrons do not escape in the first orbit through the end surface of the resonator but strike its lateral surface.

As l is increased in modes with increased resonator thickness, the injection phases of the resonance particles are shifted from the maximum of the accelerating electric field. The range of captured phases in many of the acceleration modes is greater than that in the conventional mode. The working range of acceleration modes with $l = 1.8$ and equilibrium phase $\varphi_s = 0.35$ is shown in Fig. 2.42 with indication of the lines of equal ranges of captured phases $\Delta\varphi = const$.

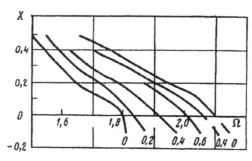

Figure 2.42. Working range of acceleration modes.

With an increase in l, however, it is difficult for resonance particles to pass through the resonator in the initial stage of acceleration. Particle trajectories in a mode with $l = 1.8$ and $x_0 = 0.3$ are shown in Fig. 2.43. It is apparent, in the first place, that the passage of particles is hindered by the cathode and, secondly, that the beam passes in the immediate vicinity of the lower right corner of the resonator.

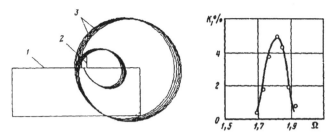

Figure 2.43. *Left*: Particle trajectories: 1) inner surface of resonator; 2) cathode; 3) particle trajectories. *Right*: Capture coefficient as a function of parameter Ω.

Despite the complexity of beam guidance in the initial acceleration region that follows from the calculation, the mode with $l = 1.8$ and $x_0 = 0.3$ was studied experimentally. The transit openings had the following dimensions: $1.2 \leq x \leq 2.1$ and $-0.18 \leq z \leq 0.18$ for the injection port, $-0.6 \leq x \leq 0.1$ and $-0.18 \leq z \leq 0.18$ for the entrance, and $-0.3 \leq x \leq 1.6$ and $-0.29 \leq z \leq 0.29$ for the exit. Here z is the axis that is perpendicular to the median plane. Note that at the first orbit the electron trajectory lies close to the corner of the accelerating cavity, but experiments have shown that the beam circumvented both the cathode and the resonator.

A measured curve of the capture coefficient as a function of the parameter Ω in the 22nd orbit is also provided in Fig. 2.43. The range of Ω variation is shifted somewhat in the direction of higher Ω values as compared with the calculations, whose results are shown in Fig. 2.42. The maximum capture coefficient is approximately the same as in the conventional second type of acceleration. In this case, the beam current rose in comparison with the conventional mode. Using a magnetron with a pulse power of 1.6 MW, the current in the 22nd orbit for a total energy of 21 MeV reached 20 mA in the conventional mode and 30 mA in the new mode.

2.4.4 Focusing Action of the Cathode Hole

In most calculations of beam dynamics in a microtron, numerical integration of the phase-radial motion of particles in the median plane is performed, but distortion which is caused by the cathode hole and cavity apertures is neglected. In fact, the size of apertures and the recession depth of the cathode are fitted empirically.

The first step in this direction was made in [2.44] in which the calculation of electrons in the vicinity of the emitter was performed by trajectograph with an electrostatic bath. When electrons are moved so far from the cathode that distortion of the cathode hole is negligible, the calculations are performed by computer. Taking into account the data which are obtained in trajectograph the three-dimensional equations of further motion are integrated numerically. The three-dimensional relativistic equations of electron motion are

$$\frac{d}{d\varphi}\left(\frac{u}{\sqrt{1-\beta^2}}\right) = -\Omega v + \varepsilon\Omega v \frac{x}{r} J_1(r)\sin\varphi$$

$$\frac{d}{d\varphi}\left(\frac{v}{\sqrt{1-\beta^2}}\right) = \Omega u - \varepsilon\Omega J_1(r)\sin\varphi \frac{ux+wz}{r} + \varepsilon\Omega J_0(r)\cos\varphi$$

$$\frac{d}{d\varphi}\left(\frac{w}{\sqrt{1-\beta^2}}\right) = \varepsilon\Omega v \frac{z}{r} J_1(r)\sin\varphi, \qquad (2.46)$$

where $r = \sqrt{x^2+y^2}$, $u = dx/d\varphi$, $v = dy/d\varphi$, $w = dz/d\varphi$, $u^2+v^2+w^2 = \beta^2$. This system of equations describes motion of electrons inside the cylindrical resonator not disturbed by holes. Because the vertical dimension of the emitter is relatively small $(\Delta z \leq 0.2)$, the vertical and horizontal motion of electrons on the initial part of the trajectory were calculated independently. For the same reason, it is possible to consider two independent criteria of stability, namely, a representative point on the phase plane (φ, Γ) has to be located inside the stable region which is restricted by the ellipse, and simultaneously the point on the plane (z, w) has to be located inside the region which is restricted by the ellipse that corresponds to the vertical aperture.

The calculations were predominantly performed for the second type of acceleration with the following values of the parameters: $l = 1.45$, $\varepsilon = 0.8$, $\Omega = 1.71$, $\overline{x_0} = 0.2$ ($\overline{x_0}$ is the coordinate of the centre of the emitter, l cavity thickness).

Figures 2.44-2.46 represent the results of calculations in the form of curves on the phase-energy plane ($\overline{\varphi}$ is the average phase of the electron in it motion through the resonator, $\Delta\gamma = (\Gamma-\Gamma_s)/\Omega$, where Γ_s is the equilibrium energy). The curves on the phase plane connect representative points which correspond to electrons emitted from the same point of the cathode (x_0, z_0) but at different initial phase φ_0.

The position of the emitter in the hole is characterized by two dimensionless parameters: recession depth δ which determines the delay of the particle in entering the non-disturbed region, and by the ratio of the recession depth to the diameter of the cathode hole (parameter $]mu$). The latter determines the strength of field on the emitter and the strength of the cathode lens.

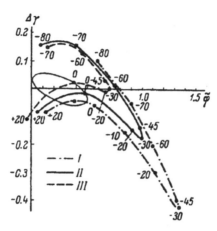

Figure 2.44. Representative curves on the phase plane in the case of non-deepened cathode ($\delta = 0$, $\mu = 0$). The numbers indicate the initial phase φ_0 in degrees; I — $x_0 = 0.115$, II — 0.200, III — 0.285.

Figure 2.44 illustrate situation with an emitter whose surface is at the level of inner surface of the resonator. However, the situation becomes more favourable when the cathode surface recedes. As it is seen from Fig. 2.45 the curves on the phase plane are brought together and this means that on average, particles have phase oscillations of lower amplitude and the amplitude of vertical oscillations is simultaneously decreased.

It is clear from the figures how the character of the curves changes with recession of the emitter. The curves on the phase plane have a sharp ridge where representative points are 'crowded' (at equidistant step of initial phase). This means that when this ridge comes to the ellipse the fraction of the electrons captured into acceleration increases. In the case of using a cone cathode hole (see Fig. 2.46) the ridge comes to ellipse and efficiency of acceleration is also increased.

2.4.5 Account of the Cavity Field Distortion by Apertures

Calculations of acceleration modes in a microtron commonly consist of numerical integration of the equations of electron motion in an undisturbed field inside the cavity (outside the cavity the

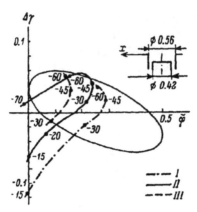

Figure 2.45.   Accelerating regime with a large capture coefficient
($\delta = 0.08$, $\mu = 0.14$). $I — x_0 = 0.03$, $II — 0.20$, $III — 0.37$.

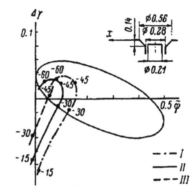

Figure 2.46. Phase curves in the case of a cone cathode hole $\delta = 0.1$). I
$— x_0 = 0.115$, $II — 0.2000$, $III — 0.285$.

trajectories are considered to be circular).   According to the
results obtained, the accelerating cavity is designed, and in most
cases experiments and calculations are in rather good agreement.
However, in some cases this is not so. For example, experiments
show that there is intense bombardment of the external side of
the cathode by electrons after their pass through the additional
aperture on the resonator's wall. Experimentally this undesirable
back bombardment of the cathode by electrons is eliminated by
screening the cathode by a tantalum damper. No calculations show
any possibility for the electron trajectory to touch the outside part
of the emitter. Certainly, this demonstrates a disadvantage of the
calculations, and it is likely that this is connected with the
distortion of the field in the cavity by its beam apertures.

Thus, for the practical use of a microtron the paraxial approximation is inadequate because the paraxial electrons are only a small part of the total accelerated beam and, strictly speaking, their motion cannot be considered as the criteria for choosing optimal microtron parameters. In order to take into account all electrons one should use a more correct description of the disturbed field in the cavity with flight apertures.

The first step in taking into account the disturbance caused by the apertures in the cavity was made by Semenov [2.45] whose results we discuss below.

2.4.5.1 *Description of the Cavity Field, Disturbed by Narrow Diametrical Slits*

For a simple description of the disturbed field in the accelerating cavity with narrow diametrical slits let us consider the field distribution across the slit is the same as the wave field in space separated by a thin conductive plane with an infinitely long slit. We will search for the solution of this problem to be a propagating wave, so that the electric field component of the wave at one side of the plane at infinity has only a normal component of electric field E_z, and at the other side of it at infinity reduces to zero. The conductive plane with the infinitely long slit can be considered as a hyperbolic cylinder

$$\frac{y^2}{1-\kappa^2} - \frac{z^2}{\kappa^2} = a^2 \tag{2.47}$$

at limit $\kappa \to 0$ in the system of degenerate elliptical coordinates (s, κ, x) being transformed to the grid coordinates (x, y, z) with rules

$$y = a\sqrt{(1+s^2)(1-\kappa^2)}, \quad z = as\kappa \tag{2.48}$$

where $2a$ is the width of the slit.

The field with such an asymptotic cannot have an x-component of electric field from symmetry demands. In this case all components of the electromagnetic field can be expressed through an auxiliary, function V satisfying the differential equation

$$\frac{\partial^2 V}{\partial x^2} + \frac{\sqrt{(1+s^2)(1-\kappa^2)}}{a^2(s^2+\kappa^2)} \times$$

$$\left[\frac{\partial}{\partial s}\left(\sqrt{\frac{1+s^2}{1-\kappa^2}}\frac{\partial V}{\partial s}\right) + \frac{\partial}{\partial \kappa}\left(\sqrt{\frac{1+s^2}{1-\kappa^2}}\frac{\partial V}{\partial s}\right) \right] + k^2 V = 0. \tag{2.49}$$

This equation has the exact solution that agrees with the an asymptotic conditions mentioned above

$$V = \frac{i}{k}e^{\pm ikx}\frac{y}{2}\left[1 - \frac{s}{\sqrt{1+s^2}}\right].$$ (2.50)

Then the approximate solution for fields in the cylindrical cavity with the diametrical slit in the plane wall can be obtained, if the field distribution across the slit considered to be the same as the distribution in the model wave. For this purpose we express the plane wave (2.50) as a series of Bessel functions and only the first term $J_0(kr)$ is kept. Then the components of the field in the cylindrical coordinate system (r, φ, z) are the following

$$E_z = J_0(gr)\frac{1}{2}\left[1 - \frac{s\sqrt{1+s^2}}{\kappa^2 + s^2}\right], \quad E_r = J_0(gr)\frac{\kappa\sqrt{1-\kappa^2}}{2(\kappa^2 + s^2)},$$

$$H_z = iJ_1(gr)\frac{\kappa\sqrt{1-\kappa^2}}{2(\kappa^2 + s^2)}, \quad H_\varphi = -iJ_1(gr)\frac{1}{2}\left[1 - \frac{s\sqrt{1+s^2}}{\kappa^2 + s^2}\right].$$ (2.51)

These equations (in the case $g = k = 2.4048/R$, R is the radius of the cavity) automatically satisfy the boundary condition on the cylindrical surface of the cavity and in limit $a \to 0$ turn continually into the expressions for the field of the undisturbed cavity

$$E_z = J_0(gr), \quad H_\varphi = -iJ_1(gr).$$

It is clear that the difference between the approximate and precise solution in the case of a disturbed cavity, as it is obvious, is determined by the half-width of the slit a. And if the condition $a \ll R$ holds, then the relative error of field distribution being expressed by equations (2.50) can be evaluated as $\delta \sim (J_0(ga)) \approx (ga/2)^2$ (for commonly used slits it is about 1%).

2.4.5.2   Condition of Resonance Acceleration

If electrons travel through an undisturbed cavity, the condition of resonance acceleration is

$$\varepsilon = \frac{1}{\Omega}\frac{eE_0}{mc\omega} = \frac{1}{2\cos(\varphi_s)\sin(kl/2)},$$ (2.52)

where $\varepsilon = E_0/H_s$ is the dimensionless amplitude of electric field in the cavity, and H_s is the guiding magnetic field in the microtron.

In case of the field in a cylindrical cavity disturbed by diametrikal slits the expressions (2.51) should be used. It is to be noted that if the condition $l \gg a', a''$ is observed (where l is the accelerating gap and a' is the entrance slit, and a'' is the exit slit) the field disturbances can be considered as provided by each slit on corresponding side of the middle plane of the cavity independently of the other slit, and the distribution of the field in the middle plane is the same as in the case of an undisturbed cavity. Then the condition of the resonance acceleration ($\Delta\Gamma = \Omega$, where Ω is the magnetic field parameter) will be the following

$$\varepsilon = \frac{1}{G(ka'')\sin(\varphi_s + kl/2) - G(ka')\sin(\varphi_s - kl/2)}, \qquad (2.53)$$

where $G(z) = zK_1(z)$, $K_1(z)$ is the Macdonald function.

2.4.5.3   The Surface of Phase Capture

To reveal the influence of the field disturbances upon the parameters of the mode with the highest capture coefficient, where the mode was used $l = 1.07$, $x_0 = 1.7$, $\Omega = 1.2$ in the first type of acceleration with injection in negative phases.

Numerical integration of the equations of electron motion was carried out up to the 17th orbit. In dimensionless form the equations of motion through the microtron cavity are the following

$$\frac{d\vec{\beta}}{d\varphi} = \frac{\Omega}{\Gamma}\left\{\vec{\varepsilon} - \vec{\beta}(\vec{\beta}\vec{\varepsilon}) + [\vec{\beta}\vec{h}]\right\}, \quad \frac{d\Gamma}{\varphi} = \Omega(\vec{\beta}\vec{\varepsilon}), \quad \varphi = \omega t, \qquad (2.54)$$

where $\vec{\varepsilon} = \vec{E}/H_l$, $\vec{h} = \vec{H}/H_s$. It was assumed that the first flight of injected electrons to the middle plane of the cavity took place in the undisturbed field. Outside the cavity, at $|z| > l/2 + 3a$, i.e. beyond the scope of the field sagging regions into the flight apertures, the electron trajectory was considered to be circular.

The results of calculations of the coefficient of phase capture $k_\varphi(x_0, \Omega)$ for a cavity with an undisturbed field and with permeable walls for electrons are shown in Fig. 2.47(a). In Fig. 2.47(b) one can see the results of the same calculations to be carried out for the case of a cavity disturbed by flight slits. Thus, it can be seen that in the case of a disturbed field the topology of the surface of capture coefficient is different from that for the undisturbed cavity, the maximum capture coefficients being displaced in the region of larger cathode coordinates x_0 and lower field parameters Ω.

Previously, in searching for the microtron effective regimes with injection from the inner wall of the accelerating cavity, the concept of phase-stability-ellipse was suggested in [2.46] to reduce

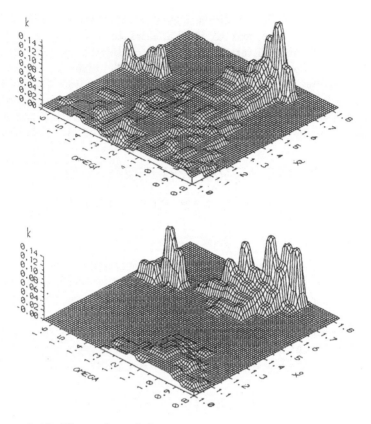

Figure 2.47. The surface of phase capture coefficient $k_\varphi(x_0, \Omega)$: (a) — in the case of unperturbed cavity with permeable walls for electrons; (b) — in the case of perturbed cavity with the flight holes in the shape of long slits of widths $a' = 0.21$ and $a'' = 0.12$.

the calculation time, and then it was used in all the following calculations. The electrons were considered to be captured into acceleration process (up to the last orbit) if they fell within the ellipse of phase stability on the third orbit. In order to determine the ellipse position in the phase plane for the optimal synchronous phase the phase oscillations were investigated only on the remote orbits, where curvature of the electron trajectory is negligible and the electron velocity differs only slightly from the velocity of light. Actually, such an approach gives a true picture of phase motion, but the possibilities of modern computers widen significantly numerical simulations of physical processes, and, in particular, make it easy to calculate phase motion of electrons up to 20 orbits.

It is necessary to emphasize that electron trajectories inside the cavity especially on the several nearest orbits are curvilinear. The changing of the trajectory curvature from one orbit to another removes the possibility of stable .acceleration when the electron passes through the middle plane of the cavity in the same phase. Stable acceleration $\Delta\Gamma_n = \Omega$) is now determined by the functional dependence $\varphi_{sn} = \varphi_s(n)$, which follows from the equation

$$1 - \int_{-kR_{n-1}\alpha_{n-1}}^{0} f(\varphi_{sn-1}, R_{n-1})dx - \int_{0}^{kR_n\alpha_n} f(\varphi_{sn}, R_n)dx = 0,$$

$$f(\varphi_{sn}, R_n) = \varepsilon J_0\left(kR_n\left(1 - \cos\frac{x}{kR_n}\right)\right)\cos\left(\frac{x}{kR_n}\cos(\varphi_{sn} + \frac{x}{\beta_n})\right),$$

$$\varepsilon = \frac{1}{2\cos\varphi_{s\infty}\sin(kl/2)}, \quad R_n = \frac{(n+1)\lambda}{2\pi}, \qquad (2.55)$$

where $\sin(\alpha_n) = l/2R_n$, R_n is the radius of the nth orbit. Here it is taken into account that microtron orbits are curviliniar inside the cavity, that the electric field depends on the transit coordinate of a particle through the cavity, and the fact that the velocity of electrons differs from the velocity of light. The expression for the dependence of electron velocity on orbit radius used is

$$\beta_n = \sqrt{1 + \frac{1}{(\Omega kR_n)^2}}. \qquad (2.56)$$

Since an ellipse of phase stability in the case of internal injection (from the inner wall of the cavity) is occupied with particles on the first orbit in a very irregular way, then the following drift of equilibrium phase leads to the dependence of the particle losses in the considered mode on the initial distribution of represented points inside the phase ellipse. For this reason it is insufficient to calculate the motion of particles only to the third orbit to obtain the one value of capture coefficient . The calculations should be extended at least to the moment when the phase drift becomes negligible and then the first revolution of the represented point on the phase plane will be finished. This takes approximately 9–10 orbits.

2.4.5.4 Focusing of Electron Beam on the Injection Half-Revolution

In the matrix theory of vertical stability the influence of fields around flight apertures on electron motion is considered to be located in the aperture's plane. Whithin the scope of such an approach it is not possible to consider injection half-revolution, where electrons run inside the cavity, and their motion is not subject to influence of apertures. Moreover, in the first type of acceleration the beam is subjected to strong defocusing of the magnetic rf field, as well as to defocusing by the exit aperture. As a result of defocusing, the electron beam is greately resricted by the entrance aperture.

However the vertical losses can be reduced by using, on the injection half-revolution, the focusing influence of the disturbed electric field on the electron beam at its turning point inside the cavity. In this way the dimensions of the entrance slit should be chosen to compensate vertical defocusing of the beam by rf magnetic field.

2.4.6 Influence of the Third Harmonic on Dynamics of Electrons

In some cases an accelerated electron beam can excite the mode E_{011} in the cavity in addition to the fundamental mode E_{010} (electric and magnetic field distribution of mode E_{011} is shown in Fig. 2.48). This oscillatory mode can be excited by the third harmonics of the beam at corresponding cavity size.

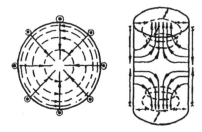

Figure 2.48. Electric and magnetic field distribution of E_{011} mode in a cylindrical cavity.

In fact, the resonance frequency of the fundamental mode corresponds to the resonance wavelength

$$\lambda_{010} = 1.305D, \tag{2.57}$$

where D is the diameter of the cylindrical cavity, and the frequency does not depend on its height l. On the other hand, the resonance wavelength of the E_{011} mode is equal to

$$\lambda_{011} = \frac{1}{\sqrt{\frac{(0.766/D)^2+(1}{2l)^2}}}.$$
(2.58)

Therefore, if the resonance wavelength of fundamental mode equals 10 cm (the diameter of the cavity is 7.66 cm), then the third harmonics of this resonance wavelength would correspond to the height of the cavity $l \simeq 1.8$ cm which corresponds to $l = 1.1$ expressed in dimensionless units. Therefore, we see that the third harmonics can really be excited by the beam in the first acceleration regime. This situation was considered theoretically [2.47] as well as experimentally [2.48] by a research group from the Lebedev Physical Institute (Moscow).

It was observed that besides the fundamental frequency there are oscillations on the third harmonic in the spectrum of the rf signal from the accelerating cavity. Figure 2.49*a* shows the dependence of the amplitude of rf oscillations (in relative units), at fundamental and third harmonic, on the magnitude of the accelerating electron current.

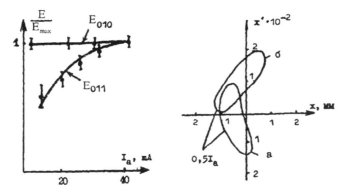

Figure 2.49. (*a*) — Amplitudes of E_{010} and E_{011} modes of rf oscillations in the accelerating cavity depending on the accelerating current. (*b*) — Momentary radial emittances of the electron beam at different time intervals from the beginning of acceleration $t = 1\,\mu$s (*a*) and $t = 3\,\mu$s.

It is seen that the amplitude of the fundamental mode is practically constant whereas the amplitude of the third harmonic depends on the magnitude of current. This fact clearly demonstrates that these oscillations are excited by accelerated electrons themselves.

The authors also observed that the radial emittance of an electron beam is different at different moments of time during the pulse (see Fig. 2.49b). These changes correlate well with changes in the amplitude of the accelerating electric field.

In addition, it was measured that a coefficient of electron capture into acceleration depends on a magnitude of accelerating beam. Influence of the amplitude of accelerating field on coefficient of capture and on a position of radial emittance of the accelerating beam can be connected with excitation of the third harmonic in accelerating cavity.

Detailed calculations [2.47] of particle dynamics in a microtron taking into account the influence of the third harmonic showed that the characteristics of an electron beam depend on the amplitude of the fundamental mode as well as on the amplitude of the third harmonic. In particular, the calculation allows us to explain the dependence of capture coefficient on the emission current which is shown in Fig. 2.50.

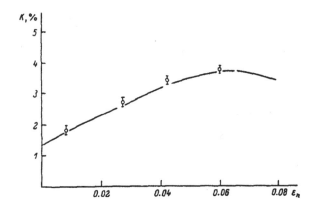

Figure 2.50. Coefficient of capture K depending on the intensity of the third harmonic of accelerating field: circle — experimental data, solid line — calculation.

The appearance of the observed dependence can be understood in the following way. First, the region of stable phase oscillations increases with an increase of the amplitude ε_n of the third harmonic. Secondly, with increasing ε_n electrons with larger initial values of the vertical coordinate are captured into acceleration. Such an increase of the effective size of an emitter is connected with the fact that there are transversal components of the electric field in the induced mode E_{011}, and these components have an impact focusing action on electrons in their flight through the cavity.

Calculations also clarified the situation with the dependence of the angle x' between the axis y and the horizontal component of the beam velocity on amplitude of the fundamental and the third harmonic. An increase of the intensity ε of the amplitude of the fundamental mode leads to an essential change of the angle x' and to an increase of the angle divergence $\Delta x'$. The dependencies of $x'(\varepsilon)$ for different values of ε_n are shown in Fig. 2.51.

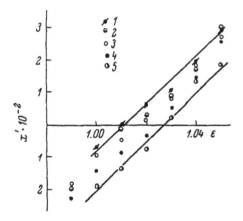

Figure 2.51. Angle between axis y and horizontal component of the beam velocity depending on the intensity of the fundamental harmonic at different values of the amplitude of the third harmonic ε_n: 1 — 0, 2 — 0.02, 3 — 0.04, 4 — 0.06, 5 — 0.08 (from [2.47]).

It follows from Fig. 2.51 that an increase of ε_n does not change the angular divergence but increases the angle x'. Such a dependence reflects the fact that an increase of ε and ε_n results in increasing transversal forces of the apertures as well as of the cavity, and these lead to increasing angular divergence and cause shift of the centres of orbits and therefore cause a change of the angle x'.

The calculations showed that value of the beam emittance remains constant at all values of ε and ε_n; only its shape and position are changed.

2.4.7 Two Frequency Microtron

The efficiency of high current microtrons is very much dependent on the amount of injected electrons that cannot be stably accelerated. To enhance the relative number of useful particles one has, as extensively as possible, to avoid injecting electrons outside the acceptance area. As it was shown by Rosander [2.49], a feasible way could be to enlarge this area with the aid of a second resonator.

In a normal microtron the electrons are accelerated in the microwave cavity by a voltage which varies sinusoidally with time:

$$V(\varphi) = \hat{V} \sin \varphi, \quad \varphi = \omega t. \tag{2.59}$$

The motion of the electrons in the phase plane depends on the form of the sine curve and is described by

$$\varphi_m = \varphi_{m-1} + 2\pi\nu\delta_{m-1}$$
$$\delta_m = \delta_{m-1} + (\sin \varphi_m / \sin \varphi_r) - 1, \tag{2.60}$$

where the coordinates, φ_m and δ_m of the mth orbit are the phase angle at which the particle passes the centre of the cavity, and the energy error over the resonant energy gain, while φ_r is the resonant phase angle and the integer ν is the resonant difference in flight time between two successive orbits measured in units of the acceleration period.

This is not very favourable since the time variation of the sinusoidal voltage is too rapid to permit stable acceleration except in the vicinity of the voltage peak. However, if an auxiliary resonator, working at some harmonic of the acceleration frequency, is put close to the main resonator, the situation will be changed. The electrons will now see a resulting accelerating voltage which can be written:

$$V(\varphi) = \hat{V}[\sin \varphi + k \sin(n\varphi + \psi)], \quad m = 2, 3, 4, \dots \tag{2.61}$$

This is a form of Fourier synthesis and by a proper choice of frequencies, amplitudes and phase difference it is possible to obtain an accelerating voltage with such a time variation that the phase-stable area can be considerably increased. The synchrotron oscillations will now follow:

$$\varphi_m = \varphi_{m-1} + 2\pi\nu\delta_{m-1}$$
$$\delta_m = \delta_{m-1} + \frac{\sin \varphi_m + k \sin(n\varphi_m + \psi)}{\sin \varphi_r + k \sin(n\varphi_r + \psi)} - 1, \tag{2.61}$$

where all φ's are referred to the main resonator. Generalizing the results of Kolomensky [1.10], one finds that phase stability for small oscillations exists as long as:

$$-2/(\pi\nu) < dV(\varphi_r)/V(\varphi_r)d\varphi_r < 0. \tag{2.63}$$

Here that means:

$$-2/\pi\nu < \frac{\cos\varphi_r + nk\cos(n\varphi_r + \psi)}{\sin\varphi_r + k\sin(n\varphi_r + \psi)} < 0. \qquad (2.64)$$

(Numerical example: $n = 2$, $k = 0.25$ and $\psi = 60°$ gives $60° < \varphi_r < 146°$. In the case of a single cavity the result is $90° < \varphi_r < 122.5°$).

Calculations of the stability areas using eqs. (2.62) have been performed for various combinations of harmonic number, n, relative amplitude, k, phase difference, ψ), and resonant angle, φ_r. One characteristic result is shown in Fig. 2.52 with $n = 3$. For comparison in Fig. 2.52 the phase stable area for a resonant phase angle of 108° when a single cavity is used is also shown.

The shown areas are valid for electrons travelling at the velocity of light. If the velocity is lower, as in the first few orbits of acceleration, alterations will occur mainly because of the finite distance between the resonator centres. Care has to be taken as this effect, determined by the actual design, could cause difficulties.

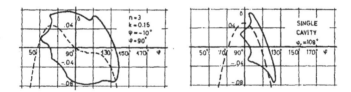

Figure 2.52. *Left*: Accelerating voltage and corresponding phase-stable area obtained with the auxiliary resonator oscillating at the third harmonic. *Right*: The same with one resonator.

Generally the increase in acceptance is paid by a deceleration in the auxiliary resonator. The fact that the auxiliary cavity, on average, is decelerating the electrons means that the circulating current itself will induce oscillations there. Probably only a smaller outer drive power is needed for correction of phase and amplitude to ensure stable operation of the accelerator.

In a machine with circular orbits it might be problematic to include an extra resonator without intersecting the first orbit. This is not the case in a racetrack microtron. In such a machine, where phase errors always occur due to the magnetic fringing fields, an enlarged phase acceptance could be of special interest.

2.4.8 Short Pulse Production

To obtain short electron pulses from a circular microtron one can shorten the length of an rf pulse but in reality it is impossible to obtain in this way an accelerated electron pulse with a duration less than 150-200 ns. Besides, the efficiency of the accelerator decreases as the time that is needed for exciting accelerating cavity is 0.3-$0.5\,\mu$s.

If the modulator, which supplies high-voltage pulses to an rf generator (magnetron or klystron), allows for changes in the pulse length then one can obtain short pulses of accelerated electrons by shortening the modulator pulses. For example, ~ 50 ns pulses from the electron were obtained at the 30 MeV microtron in Obninsk [2.50]. Figure 2.53 shows oscillograms of the accelerated and injected currents obtained in this mode of operation. Certainly, the efficiency of the microtron is lowered as part of the rf power is lost for excitation of the resonator.

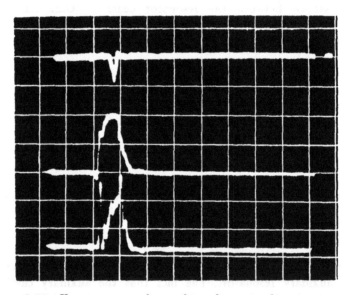

Figure 2.53. *Upper curve* — the accelerated current, *low curve* — current which is injected by the cathode.

A more efficient and simpler way to shorten the beam duration is probably to place a small inductive magnet inside a microtron. With such a magnet one can easily synchronise its operation with the main modulator pulse. The main advantage of this mode of operation is that such a bending magnet, being placed at the last orbit, does not practically change the load of the accelerating cavity by the accelerated beam.

In a race-track microtron the situation is more favourable. As it was experimentally shown by Rosander [2.51] in a race-track microtron electron pulses of a duration much shorter than the time constant of the linear accelerator can be accelerated. As the pulses are short, part of the rf energy stored in the linac may be utilized so that the instantaneous beam power can well exceed the rf drive.

In these experiments, electron pulses of about 45ns duration and 30keV energy were injected from the diode electron gun into the linear accelerator. The linac was fed by rf pulses sufficiently long to ensure that the microwave fields had reached steady conditions before the injection took place. After acceleration through 15 orbits to 50 MeV the extracted electron pulses were about 25 ns (fwhm) long and almost triangular in shape as shown in Fig. 2.54. The figure also contains the central part of a pick-up signal from the linac showing how the rf fields sag during acceleration of the heavy load and the recovery of the fields after the electrons have left the accelerator.

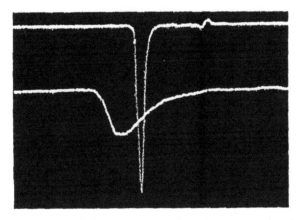

Figure 2.54. Oscillogram demonstrating short-pulse acceleration to 50 MeV. *Upper trace*: electron current, *lower trace*: linac voltage. As the electron current is measured outside the microtron this pulse is delayed as compared to the linac response.

The attained peak beam power was 3 MW, while the klystron output was 1.9 MW. This power, though partly spent on waveguide losses and mismatch reflections, is much higher than what is needed for driving the accelerating voltage to the upper limit of the interval of stable phases. The electrons in the initial part of the injection pulse therefore experience a linac voltage which is too high and get lost. Nevertheless, they load the linac so that the voltage decreases and thus the rest of the electrons can be stably accelerated.

The electron beam formed by a microtron in the repetitive pulse mode consists of a sequence of packets of electron bunches, following one another with the frequency of the pulse modulation of the accelerating field. In some cases only single packets of bunches are required for experiments. Separation of one packet from the extracted beam is simple by employing a pulsed deflecting magnet, as was performed by Zav'yalov and Semenov [2.52].

The magnet enables a selected electron packet to pass into the working channel, without exposing it to the magnetic field, while the rest of the beam is dumped onto a buffer target. The current pulses feeding the magnet are synchronized with the microtron current pulses through a clearance-to-trigger logical circuit. A clearance-to-trigger signal for the magnet is not generated in order to pass the required packet into the transport channel. This method of commutation makes it possible to reduce substantially the requirements both for field uniformity in the magnet and the form of the current pulse feeding it.

The construction of the magnet is shown in Fig. 2.55. Two flat, single-layer coils are fastened 12 mm from one another on dielectric supports and are placed inside a stainless steel vacuum chamber. The coils are made of copper tubing 4 mm in diameter with a wall 0.5 mm thick; in order to give it an oval shape the tubing was annealed and rolled between rollers with a 2 mm gap. After this the tube was wound on an expander with the required shape. In order for the deformation of the tube to be uniform, hydrostatic oil pressure equal to 50 atm was created inside it when it was bent. After a second annealing and chemical cleaning, strips of paper 0.2 mm thick were laid between the loops and the coils were glued with epoxy resin. The ends of the coils were soldered into water-cooling pipes, which simultaneously functioned as contact conductors. The coils were electrically connected in series, and in parallel along the water-cooling pipes. To create electrical decoupling, the water cooling the coil is fed through dielectric hoses.

The magnet is supplied by the sinusoidal-pulse generator and can separate individual current pulses from the electron beam of a pulsed microtron. A working field of 1.4 kG is created in the 12 mm gap between the two flat coils of the magnet.

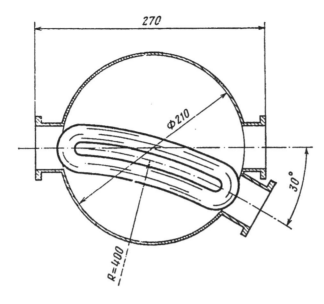

Figure 2.55. Diagram of the arrangement of the deflecting coils in the vacuum chamber of the magnet.

2.5 Continuous Wave and 3 cm Microtrons

As a rule, microtrons operate in pulsed regime in the 10 cm wavelength range. Nevertheless, there is no principal physical objection to building continuous wave microtrons (CW) and to use generators with shorter wavelengths.

Continuous wave accelerators are of special interest for nuclear physics. This interest is connected with the fact that transition from a pulse to a continuous mode of operation allows to increase an intensity which is accessible for physical experiments made with electronical methods of detection of products of nuclear reactions.

As for 3 cm microtrons, their main advantage is in their compactness. The rather small size of such machines gives an opportunity to use them as an easy transportable tool for geophysical researches or non-destructive testing in-the-field conditions.

2.5.1 Accelerating Mode with Small Energy Gain

Usually in circular microtrons a multiplicity of the acceleration at two neighbouring orbits differs by unity and therefore the minimum possible energy gain of an electron in its passage through an accelerating cavity equals

$$\Delta\Gamma_{min} \geq \Omega_{min} = 1/\nu . \qquad (2.65)$$

Here ν determines the integer number of the period of accelerating field T_0.

Actually, for $\nu = 2$ an acceleration by the cylindrical cavity is performed for Ω lying in the limits $0.7 \div 1.5$ (for the first type of acceleration). Thus, decreasing the parameter Ω is only possible by increasing the factor ν. The smaller parameter Ω, the less rf power which is necessary for producing a proper intensity of accelerating electric field in the cavity. Such regimes with small Ω are of interest for CW microtrons as well as for microtrons which are operating in 3 cm wavelength region.

The calculations of regimes with $\nu = 3$ for cylindrical cavity were performed by Rodionov and Stepanchuk [2.53]. The accelerating region for the cavity with height $l = 1.2$ and resonance phase $\varphi_s = 0.35$ is shown in Fig. 2.56. The numbers which are presented by curves in this figure correspond to the widths of initial phases of electrons, which are captured into acceleration. The maximum width corresponds to the values of l and φ_s presented above, and practically equals the width of the interval of initial phases in the common first type of acceleration at $\nu = 2$.

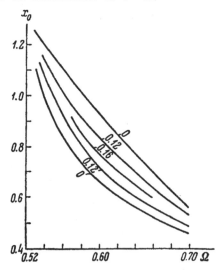

Figure 2.56. Small Ω acceleration regions for cylindrical cavity.

At different values of Ω the electrons which are captured into the acceleration have different outgoing phases from the cathode. For example, at $\Omega = 0.7$ the electrons which are captured into acceleration are emitted in the phase interval $-0.58 < \varphi_0 < -0.48$ while at $\Omega = 0.54$ the corresponding phases of injection are $-0.08 < \varphi_0 < 0.04$.

Figure 2.57 illustrates the trajectory of electrons in these regimes. At all values of 0 electrons are travelling through the cavity near its axis $(-0.3 < x < 0.3)$ and the coordinate of electrons outgoing from the cavity after their first turn equals $1.8 < x_1 < 2.4$.

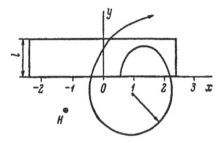

Figure 2.57. Trajectory of electrons at $\Omega = 0.66$ in cylindrical cavity.

The accelerating regimes at $\Omega = 0.54$ allow to increase the rf power needed for excitation of the cylindrical cavity by approximately 1.5 times and the electric field intensity is by 1.25 times less relative to the common first type of acceleration.

2.5.2 Non-Superconductive CW Microtron

As it was considered above, calculations show the possibility of acceleration of electrons in a microtron with small energy gain per turn. This possibility is the most essential point for normal (non-superconductive) variant of a continuous wave microtron, because CW rf generators have an average power on the level of some hundred kW.

The normal CW microtron was designed and put into operation by Slobodyanyuk with coauthors in the Institute of Physics and Energy (Obninsk, Russia) [2.54]. The nigotron [2.55] — rf generator of the magnetron type of 1607 MHz frequency — was used as a power supply. It was managed to obtain from this generator up to 150 kW of useful power, but actually the authors managed to transfer from the generator to the cavity only 30-50 kW. Unfortunately, this machine has now dismantled.

To avoid temperature drift, intensive water-cooling was introduced to maintain the temperature with accuracy $\pm 1^\circ$C. Injection of electrons was performed by an electron gun placed near the accelerating cavity. At $\Omega = 0.108$ the accelerated current on the 8th orbit was equal about $100\,\mu$A. The distribution of the current obtained is shown in Fig. 2.58.

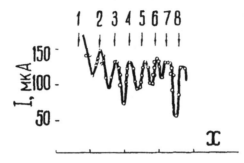

Figure 2.58. Dependence of the current in a CW microtron on the number of the orbit; acceleration was performed with $\Omega = 0.108$, injection energy of electrons $U = 7.6\,$kV (from [2.54]).

It was also shown that it was possible to accelerate electrons at $\Omega = 0.26$ which corresponds to the energy at the last orbit 1.32 MeV. The size of the accelerated beam was 6-8 mm in diameter.

It is necessary to underline that the interorbit distance in this machine was not equidistant. In fact, this is a non-relativistic accelerator, and the possibility of acceleration of electrons in the microtron mode shows that it is possible to accelerate heavy ions in the same manner.

The key to producing a continuous beam is also the use of race-track geometry with a superconducting acceleration cavity. Superconductivity reduces the losses in the cavity walls and makes it possible to maintain high radiofrequency fields continuously rather than in short pulses. This idea was successfully applied to a linear accelerator in the Illinois University microtron which we consider in the following chapter.

2.5.3   3 cm Microtrons

The majority of microtrons are operated at 10 cm wavelength. The primary choice of this wavelength region is determined by two factors:

(i)   the necessary rf equipment had been completely manufactured for military radars in 50's when microtrons began to be designed and made;

(ii) the size of an rf accelerating cavity is suitable for obtaining high intensity electric fields and for placing a thermionic cathode on its wall for injection of electrons.

On the other hand, transition to larger than 10 cm wavelengths will lead to the considerable increase in the size and weight of a microtron because the size of cylindrical accelerating cavities which are operating in E_{101} mode and the interorbit distance are proportional to the wavelength of the accelerating field. That is why there are no microtrons which operate at higher than 15 cm wavelength,

However, the development of powerful 3 cm magnetrons in the 70's led to the generation of compact microtrons [2.56, 2.57]. The main initial problem in making such a machine is obtaining a sufficient increase of electric field intensity. Actually, it was well known from experiments with 10 cm machines that at rf electric field intensity $E \simeq 600 \, kV/cm$ rf discharges take place between the copper walls of a cavity.

The only way to avoid this problem with electrical strength is to use an accelerating mode with relatively small parameter Ω which governs the process of acceleration as well as the intensity of electric field in the cavity. It was considered above that calculations showed that it is possible to decrease the value of parameter Ω by using $\nu = 3$ instead of commonly used $\nu = 2$ (integer number ν determines the number of periods of accelerating field T_0 by which a time of the revolution of neighbouring orbits are differed).

The working region of the microtron parameters in cylindrical cavity with width $l = 1.15$ at resonance phase $\varphi_s = 0.35$ is shown in Fig. 2.56. As it seen from Fig. 2.56, the effective size of the cathode in the median plane (the plane of orbits) equals $\simeq 0.15$ (at $\lambda = 3 \, cm$ it corresponds to 0.8 mm). The vertical size, as follows from calculation, can be set equal to 0.6 mm, and therefore an emitting surface is $\simeq 0.6 \, mm^2$, i.e. sufficient to obtain an emission current up to 10 mA.

The accelerating cavity was excited by a 240 kW-power 1 μs-pulse magnetron at 3.2 cm wavelength which operates at a frequency 830 Hz. The copper-made cavity had the horizontal apertures $2 \times 6 \, mm$.

Figure 2.59 shows a cross section of the magnet with the vacuum chamber used. The vacuum chamber size was 255 mm diameter and 40 mm height. The outer diameter of the magnet equals 580 mm, the maximum magnetic field intensity 4000 Oe, the total weight of magnet with chamber 300 kg.

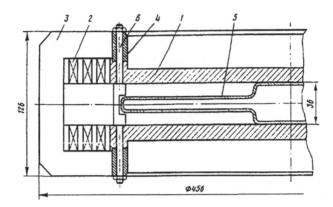

Figure 2.59. Schematic cross section of the magnet and vacuum chamber of 3 cm microtron: 1 — magnetic pole, 2 — coils, 3 — reverse magnetic core, 4 — transition ring, 5 — vacuum chamber, 6 — matching stud (from [2.57]).

Due to the small space available in a 3 cm microtron, a special construction of the cathode was designed [2.58]. The main elements of the injector unit are shown in Fig. 2.60. The injector is an LaB_6 specimen 1 in the form of a rectangular parallelepiped, which is placed in a tantalum clip 2 and has a projection 3, from whose end, which has an area of 0.7×0.7 mm, electrons are emitted into an accelerating resonator. Welded to the clip is a tantalum cross piece 4 with a holder 5 in the form of a molybdenum rod welded to a copper plate 6, which is mechanically attached to lug 7. The auxiliary cathode 8 is a stock indirectly heated end-type tungsten-barium cathode. A holder 9, which is attached to lug 10, is welded to the cylindrical housing of cathode 8. A heat shield in the form of a tantalum foil can be placed between the housing of cathode 8 and holder 9. The auxiliary cathode has an emitting-surface diameter of 2.8 mm, and the distance between the injector 1 and cathode 8 is 1.1 mm. Also shown in Fig. 2.60 are the heater lead 11 for the tungsten-barium cathode and the cover 12 of the microtron's accelerating resonator.

Experiments were performed with a 3 cm microtron in a dismountable vacuum chamber with constant evacuation. After bench measurements of its initial characteristics, the tungsten-barium cathode was placed in the emission mode by the conventional technology, and at a temperature of $T \simeq 1500$ K, the injector was degassed with gradual heating to ~ 2150 K. The injector was heated by the electron beam from the auxiliary cathode with an increase in the dc voltage between the cathode and the injector in steps of 10-30 V with holding at each voltage for 10-15 min;

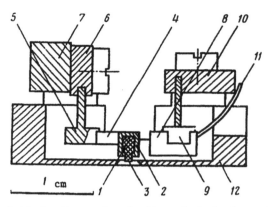

Figure 2.60. Lanthanum-hexaboride cathode for 3 cm microtron.

a pressure increase of $> 5 \times 10^{-5}$ Torr was not permitted. The injector was held at a temperature of $\simeq 2150$ K until a pressure of $< 5 \times 10^{-6}$ Torr was attained in the system. These degassing conditions prevented contamination of the tungsten-barium cathode as a result of gas liberation during electron bombardment of the injector.

Two clip designs were tested: one with a face of the LaB_6 specimen open for the electron beam and another with the face covered by a tantalum plate. A more stable current from the auxiliary cathode with the injector put into operation was obtained with the second clip version, apparently due to reduction of the amount of lanthanum-hexaboride evaporation products — primarily boron — that reached the tungsten-barium cathode.

To reduce the weight of a 3 cm microtron used as a portable accelerator in defect detection, the special pulse modulator of a magnetron was built [2.59]. The microwave generator can be moved independently around the components being tested to a distance up to 10 m from the stationary power supply and control circuit. To reduce the distortions in the output modulation pulse while it is being transmitted to the magnetron a significant distance from the modulator, the step-up pulse transformer is placed in the microwave unit of the microtron and connected to the shaping line of the modulator via a cable. To transmit the pulse without distortion, the load impedance must be matched with the wave impedance of the cable. Since the transmission time along the high-voltage cable is much less than the pulse plateau duration, matching the cable impedance during the pulse edge time is important. During the rise time of the edge before the magnetron is excited, the cable is practically unloaded; then the reduced magnetron impedance changes as the magnetron current grows, exceeding the cable's wave impedance until the current reaches its nominal value. During this time the cable output is, therefore, mismatched. The reflected wave that emerges as a result of this mismatch is absorbed by the resistor in the shaping line's correction circuit.

Kazakevich with colleagues [2.60] designed and built a special transportable 3 cm microtron for geophysical research. The external diameter of the vacuum chamber is 20 cm and it is 4 cm high. The acceleration mode used was the same as noted above ($\Omega \sim$ 0.6, $\varepsilon \simeq 1.1$). Using the magnetron with 240 kW pulse power and 1 μs pulse width, it was possible to obtain at the last 13th orbit an accelerated 4 MeV beam current of $\sim$9 mA and 0.65 μs duration.

Chapter 3

Race-Track Microtrons

The concept of a split or race-track microtron (RTM) in which two uniform-field dc magnets with parallel edges are separated by a distance large compared to the magnet dimensions, and a linear accelerator placed in the common straight section, was first suggested by Schwinger as it was reported by Schiff [3.1].

However, the problem of the defocusing effect in crossing the fringing field of the two main bending magnets was unsolved for many years, until in 1961 Canadian researchers Brannen and Froelich [3.2] in the University of Western Ontario built a 12 MeV machine utilizing four sector magnets. This design guaranteed axial stability making use of the focusing effect produced on electrons when they crossed two extra magnet edges at an angle of 45°. This machine, however, made use of a single accelerating cavity with a maximum energy gain per turn of the order of 1 MeV and so closely resembled the father circular microtron.

But the development of standing wave linear accelerators at Los Alamos [3.3] in the end of sixties permitted to realize the hope of having a greater final energy out of the machine: in 1972 in London, Canada a variable energy race-track microtron [3.4] was built which made use of a three cavity standing wave side coupled linear accelerator, which made it possible to achieve a 3 MeV energy gain per turn and a total final energy of 18 MeV. This machine used a focusing system similar to the previous RTM, but instead of two equally strong sector fields separated by a slanted field-free region, one magnet with three sectors, all with different field strengths (hills and valley), was used on each side of the accelerator.

In 1967 Babić and Sedlaček in Stokholm [3.5] proposed equipping each of the magnets with an oppositely magnetized thin magnet along the whole length facing the linac, in order that a narrow part of the fringing region gets a reversed field. In such a way the axial defocusing was turned into focusing without requiring the somewhat complicated four sector structure.

This invention stimulated the construction of several race-track microtrons with an energy gain per turn of 2 MeV or more. The three more famous machines were the 50 MeV at the Royal Institute of Technology (Stockholm) [3.6], and two 100 MeV accelerators, one in Lund [3.7] and the other in Wisconsin [2.19], these two both used as injectors of small rings. However even race-track microtrons without any focusing in the dipole magnets have been contemplated: for instance a 30 MeV microtron has been developed in Moscow [3.8] which uses quadrupole focusing items in any orbit.

3.1   Construction and Principles of Operations

Besides the book of R. E. Rand *Recirculating electron accelerators* [3.9] published in 1984, there are also some reviews devoted to race-track microtrons [3.10, 3.11, 3.12] in which different aspects of these machines are considered. The subject of this section is partially based on these publications.

The race-track microtron in its more usual (standard) design (Fig. 3.1) consists of a couple of 180° bending magnets facing each other and separated by a field free zone in which a linear standing wave accelerator is placed. The electrons, injected in the linac structure by means of an electron gun, eventually a preaccelerator and related optics, are accelerated toward a magnet.

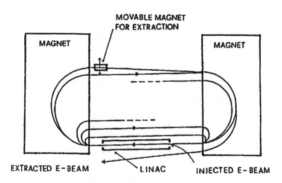

Figure 3.1. Race-track microtron layout.

Acceleration takes place in a recirculating way, as the beam is turning around in the magnets and passing through the linac several times until the final energy is reached. This gives a compact design and a short accelerating section. Recirculating a high average power beam in a small machine is not without problems. If the beam, or part of it, is lost inside the accelerator, thermal drift problems or even damage may occur. Beam losses, must therefore be kept low in the accelerator structure which also must be efficiently cooled.

Let us assume electrons are injected into the linac with a total energy of $E_{inj} = mc^2 + eV_{inj}$ where eV_{inj} is the kinetic energy. On each pass through the linac they will gain an energy ΔE. To achieve resonance and acceleration the same two conditions as for the circular microtron must be fulfilled:

- the revolution time of the first orbit must be an integral multiple, μ, of the rf period, T_0,
- each revolution time must exceed the preceding one by an integral multiple, ν, of the rf period.

Making three major approximations:

(i) magnetic fringe fields are negligible (hard edge field),
(ii) electron velocity is assumed equal to that of light, c, in all orbits,
(iii) transit time effects in the linac gaps are neglected,

we get:

$$T_1 = \frac{2l}{c} + \frac{2\pi}{eBc^2}(mc^2 + eV_{inj} + \Delta E) = \mu T_0, \qquad (3.1)$$

$$\Delta T = T_n - T_{n-1} = \frac{2\pi}{eBc^2}\Delta E = \nu T_0, \qquad (3.2)$$

where l is the field-free distance between the magnets and B the strength of the homogeneous magnetic field giving an energy gain per turn, ΔE. The two equations give the energy gain and the magnetic field:

$$\Delta E = \frac{\nu}{\mu - \nu - 2l/\lambda}(mc^2 + eV_{inj}) = \frac{\nu}{\mu - \nu - 2l/\lambda}E_{inj}, \qquad (3.3)$$

$$B = \frac{2\pi f}{ec^2}\frac{\Delta E}{\nu}, \qquad (3.4)$$

where $f = 1/T_0$, and $\lambda = cT_0$.

As the denominator of equation (3.3) can be chosen at will the energy gain per orbit is in the hands of the designer, as long as the magnetic field strength can be obtained. The phase stability conditions in an RTM is the same as in a conventional circular microtron, and therefore an accelerated beam has similar small energy spread. The fringe field of the bending magnets causes axial defocusing as the beam is already slightly bent before passing the magnet edge, so that the fringe field region will be passed at an oblique angle to the edge.

According to the proposal of Babić and Sedlaček [3.5], axial stability can be achived by using auxiliary magnets with opposite excitation near the edge of the main magnet sectors. Besides giving the electrons axial stability the auxiliary magnets correct the first orbit so that the shape of the orbit resembles the 'sharp edge' case and its length approximately satisfies the resonant condition.

In the radial plane there is no focussing, except in the linac electric fields unless extra focussing elements, like quadrupoles, are inserted along the beam trajectories.

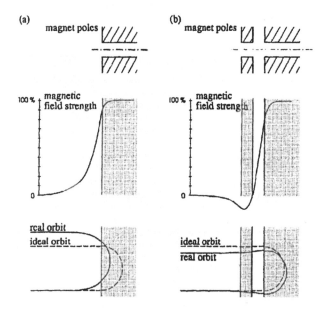

Figure 3.2. Edge focussing with auxiliary poles. (*a*) Normal defocusing magnet. (*b*) Axially focussing system with auxiliary and main magnets.

In order to extract the beam in the same output position at the desired energy a magnet is placed in the proper orbit which deflects the beam by an angle so that during the reflection by the following magnet it overcomes the common axis and exits the machine.

The important points to note are that the magnetic field strength in an RTM is proportional to the resonant energy gain as in the circular microtron but that the denominator of Eq. (3.3) can be chosen at will and can be also less than one with a proper choice of the distance between magnets. In this way the energy gain per orbit and thus the magnet field strength are in the hands of the designer because they can be chosen almost independently from the injection energy. As magnets can be made up to about 1.6 Teslas microtrons can be constructed with energy gain per turn of the order of 10 MeV in the fundamental mode which is $\nu = 1$.

Anyway this freedom in the choice of parameters is compensated by the necessity of setting at the proper value the different parameters at one time. For instance, usually, the magnetic field is of fixed value, and the distance between the magnets is tuned for the fulfillment of the previous equation: this comports a mechanical precision method which allows the movement of at least one magnet, a magnet which can weight several tons.

The maximum tolerable magnetic field variation depends on the largest allowed orbit length deviation and is usually between $10^{-3} - 10^{-4}$. So the requirements which the magnetic field must obey are stringent in terms of uniformity but nowadays a great degree of precision in the magnet design can be reached by computer simulation.

3.2 Types of Race-Track Microtrons

Race-track microtrons can be separated into three different categories.

The first category is standard pulsed RTMs in which the main magnets are simple flat-pole magnets and the Babić-Sedlaček contrapole scheme is used to prevent vertical defocusing from the fringing of the main magnets. In this type of microtron another problem arises from the short distance of the first return path from the common axis. This distance is dependent from the injection energy:

$$l = \frac{n\lambda}{\pi} \left(1 + \frac{E_{\text{inj}}}{\Delta E} \right) . \tag{3.5}$$

For small values of injection voltage respective to the first pass gain this distance is smaller than or too close to the outer cavity wall. In order to avoid this obstruction a shielded passage through the linac cavities has been constructed in RIT in Stockholm [3.6].

Another solution, invented at the same time in Wisconsin and Lund [3.7, 2.19], is to increase the injection energy by means of special recirculation into the linac by a couple of magnets placed near a big magnet. In this way the real injection energy in the race-track microtron is the sum of the injection voltage and the energy gained in the first pass through the linac, usually close to the energy gain per turn. This scheme, however requires the use of a standing wave linac which can accelerate in both directions.

The third solution is changing to a higher mode of operation: the second mode exhibits double the distance between the orbits and so allows the accommodation of several focusing items all around the vacuum pipes of the paths: this has been the solution adapted by the Lebedev Physical Institute in Moscow [3.8].

Moreover, the various items added mean that a stronger focusing machine can transport a higher current beam than all the other race-track microtrons, in which the focusing is weak. The maximum energy ranges of this type of machine range between 50 and 150 MeV, but an extraction system can be applied that allows extraction from any or selected orbits,

These are the more widely built RTMs and they are used mainly for injection in rings and medical applications, we will consider the latter in Part III.

In a 'classical' race-track microtron the bending magnets are homogeneous sector magnets and provide 180 degree bends for the successive orbits. Examples of microtrons based on uniform-magnetic-field bending magnets are the 50 MeV microtron at the Royal Institute of Technology in Stockholm [3.6], the 100 MeV Aladdin microtron in Stoughton [2.19], the 100 MeV ENEA microtron in Frascati (Italy) [3.13], the 100 MeV race-track microtron of Lund (Sweden) [3.7], the microtron cascade MAMI in Mainz (Germany) [3.14], 20 MeV microtron in the Lebedev Physical Institute (Moscow, Troitzk) [3.8].

A lot of effort has to be made to obtain good homogeneity of the magnetic field of flat magnet poles [3.14]. The weak focusing in these machines is in essence obtained by the cavity, in combination with reversed field clamps [3.14], solenoids and auxiliary magnets in the return paths of the orbits [3.8]. Due to this weak focusing, the trajectories are sensitive to magnetic field imperfections and alignment errors [3.7]. This is the main disadvantage of these relatively simply designed and manufactured machines.

In the second class of pulsed RTMs three or four sector bending magnets are used. In these machines the magnetic field provides strong focusing and, simultaneously, synchronism is maintained. Such a microtron is less sensitive to magnetic field imperfections and alignment errors. This principle was used for the design of the two Endhoven (The Netherland's) microtrons for a free electron laser (25 MeV) and for injection into a storage ring (75 MeV). The bending magnets of the race-track microtron in Eindhoven contain two distinct sectors with different magnetic field strengths (this is achieved by different air gaps which are 20 mm and 17 mm, respectively) and reversed field clamps which provide additional axial focusing. This kind of magnet provides focusing forces at three edges: at the entrance of the magnet, at the exit of the magnet and at the transition of the sector boundary.

The basic components of the 75 MeV RTM in Eindhoven [3.15] is shown in Fig. 3.3. This accelerator serves as an injector for a 400 MeV storage ring.

With these two sector magnets, it is no longer possible to obtain a pure 180 degree bend for the electrons. As a consequence, it is impossible to have simultaneous horizontal and vertical stability in a race-track microtron with parallel azimuthally varying field magnets, as it was shown by Delhez [3.16]. Simultaneous stability can be achieved only by rotating the magnets through the median plane over a small tilt angle. The electrons are injected at a kinetic energy of 10 MeV. On each passage of the accelerating structure, they gain an energy of 5 MeV, and after 13 successive turns the electrons are extracted.

An array of small correction magnets are placed on the symmetry axis, in between the bending magnets. These correction magnets are used to compensate for beam displacements in the bending plane caused by factors such as magnetic field imperfections and misalignment of the bending magnets.

The third class of RTMs is continuous wave machines, to which the following section is devoted.

3.3 Continuous Wave RTMs

In the case of race-track microtrons there is more space available than in circular machines, and one can get a CW beam by using a superconducting or a normal conducting linear accelerator. At first sight the use of a normal conducting linac seems less natural for a CW machine as the necessary rf power becomes quite high, but the refrigerating system for a superconducting structure does not need very much less power, is rather complicated and needs

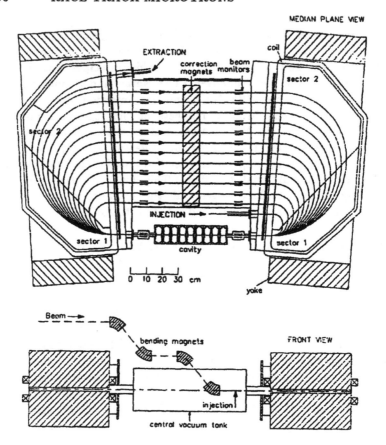

Figure 3.3. Median plane view and front view of the Eindhoven race-track microtron.

extra manpower, so superconductivity does not save any money in this case. Nevertheless, both these possibilities have been realised.

3.3.1   Race-Track Microtron Using a Superconducting Linac

The only RTM using a superconducting linac (MUSL-I) was built in the University of Illinois [3.17] and operated for about 3 years in the 70's. The duty factor was limited to about 50% by thermal effects. The system could supply $5\,\mu A$ electrons with energies up to 19 MeV and was successfully operated on a 24 hour per day schedule for nuclear physics experiments.

Electrons of 270 keV were chopped, bunched, and deflected onto the linac axis where they were accelerated to 0.7 MeV by the $\frac{3}{2}\lambda$ section and to 3.5 MeV by the $1\frac{3}{2}\lambda$ section. Because the effective phase difference between the two sections was about 70 degrees for relativistic electrons, little energy was contributed to the recirculated beam by the $\frac{3}{2}\lambda$ section and the energy gain on the subsequent passes was only 3.1 MeV. The electrons were returned to the linac by uniform field magnets whose pole pieces were 114 cm wide and 56 cm deep. The active magnetic clamps extended along the entire entrance edge of the turn-around magnets and provided a reverse field of about 10 percent of the main field to compensate for the defocusing of the fringe fields, as well as some additional vertical focusing. There were a number of cylindrical lenses made up of uniform field triplets which were used for adjusting the vertical focusing on the separate return paths.

All the magnets in the system were simple rectangles which had a focusing effect only in their vertical direction. The horizontal focusing took place only on the common linac axis by means of the linac fields and by two quadrupole singlets which were on this axis.

The phases of the returning first, second, and third pass beams were adjusted by means of the bypass angle, the magnetic field in the end magnets, and the spacing between them. The phases of the fourth and fifth pass beams needed only small corrections, which were made with pole face currents that affect the magnetic fields in strips covering only the outer portions of their semicircular paths.

The beam, after 3, 4, 5, or 6 passes through the linac could be deflected by 17 degrees into a common channel that led to the two 36.5 degree rectangular magnets which bent the beam into two experimental areas. In one of these areas a GeLi detector was used to study resonance fluorescence from nuclei at energies below their neutron thresholds. The second experimental area contained the tagged photon facility usually referred to as the photon monochromator. The tagged photon beams were used to study elastic photon scattering, photoneutron cross section, and photofission.

The $\frac{3}{2}\lambda$ and the $1\frac{3}{2}\lambda$ niobium sections have operated reliably without dismantling the cryostat and without any major difficulties for about three years. These sections which were processed only by chemical polishing had initial Q values above 10^8. These were operated at 4.2 K with energy gains of 2 MeV per meter and would require 100 watts of cooling at that temperature if operated continuously.

MUSL-I was dismantled in 1977 and a cascade microtron

MUSL-3 was to be build. The first stage of this machine MUSL-2, producing 80 MeV electrons, was in operation for many years. Many components from MUSL-I were used in MUSL-2 but the injected beam was obtained from a 3 MeV van de Graaff accelerator and the original linac was replaced by 6 m niobium structure which was capable of operating CW with energy gain up to 13 MeV using only 10 W of microwave power at temperature of 2 K. The RF power needed was relatively small due to the high Q value of the structure equal to 3×10^9. Unfortunately the high-energy machine was not built.

3.3.2 Normal Conducting RTM MAMI

At the University of Mainz an electron accelerator with a maximum energy of 855 MeV and 100% duty cycle has been built. The heart of this accelerator is a big race-track microtron which has given the name MAMI (standing for Mainz Microtron) to the whole project. A first two stage version of the machine, known as MAMI A, operated successfully from 1983 until 1987 with a maximum energy of 187 MeV and a beam current of 65 μA [3.14, 3.18]. Since the end of 1987 the full machine, sometimes called MAMI B, has been assembled and in a new building and was put into operation in 1990.

The MAMI accelerator consists of a linear accelerator as injector and a cascade of three microtrons using normal conducting RF structures (see Fig. 3.4).

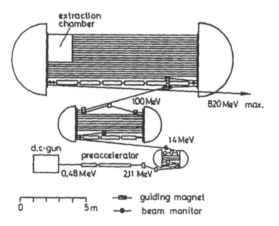

Figure 3.4. The cascade of three race-track microtrons with the injector linac.

In this way a continuous wave operation becomes possible for the first time at medium high energies ($E_{\max} \simeq 855\,\text{MeV}$) and high beam currents ($I \simeq 100\,\mu\text{A}$). The energy may be varied from 180 MeV to 855 MeV in steps of 15 MeV by kicking the beam off the separated tracks into the extraction channel. The beam will be distinguished by an excellent quality in both the transverse and longitudinal phase space.

The theory behind the MAMI design is immediately seen from the basic linac equation

$$(\Delta E)^2 = rLP\,, \tag{3.6}$$

where ΔE is the energy gain, r is the shunt impedance per unit length, L is the linac length and P is the total RF power. Since in normal conducting structures r is of the order of $50\,\text{M}\Omega/\text{m}$, a δE of the order of several 100 MeV requires such a large product $L \times P$ that only pulsed operation is technically feasible. By recirculating the beam n times, however, the above equation becomes

$$E^2 = (n\Delta E)^2 = rLPn^2\,. \tag{3.7}$$

Thus, recirculating a beam 30 times saves three orders of magnitude in $L \times P$, making a normal conducting structure a convenient and economic device of proven toughness and reliability also for CW operation. Clearly, such a large number of recirculations requires a simple and economic recirculation scheme, as given by the race-track microtron. This scheme is non-isochronous, implying that a resonance condition has to be fulfilled between energy gain per pass and magnetic flux density in the reversing magnets such that the path length has to increase from one linac passage to the next by an integer number ν of wavelengths (1 or 2, in practice). In the case of hard-edge magnets and $v = c$ this condition becomes as

$$2.096\Delta E = \nu\lambda B\,, \tag{3.8}$$

where ΔE is the energy gain in the linac in MeV, λ the rf wavelength in cm and B the flux density in the magnets in Tesla.

Acceleration occurs off the crest of the rf wave and the particles oscillate about a certain stable phase angle at which the energy gain is equal to its resonant value ΔE. The longitudinal bucket size is strongly dependent on this phase angle and is drastically reduced with increasing ν.

By a rule of thumb in any microtron the energy at the first magnet passage should not be significantly smaller than one tenth of the output energy. Thus, a specific problem of normal conducting CW microtrons is obtaining a beam of sufficient energy at injection since, unlike with pulsed microtrons, the microtron linac itself cannot be used as a preaccelerator because of its low field gradient.

Nuclear physics experiments were started at MAMI in 1985. The main investigations were directed to coincidence experiments with electrons, photonuclear reactions with real photons, and experiments with polarized electrons, etc. Operation experience with this machine shows the very high quality and stability of the beam and high level experimental data were obtained.

3.3.3 CW RTM of the Moscow University

A continuous wave race-track microtron with the maximum energy of 175 MeV and beam current 100 μA is under construction at the Institute of Nuclear Physics of Moscow State University [3.19]. To a certain extent this project has common features with the MAMI-A [3.14] and NIST-Los Alamos [3.20] machines. The plan view of the RTM is shown in Fig. 3.5.

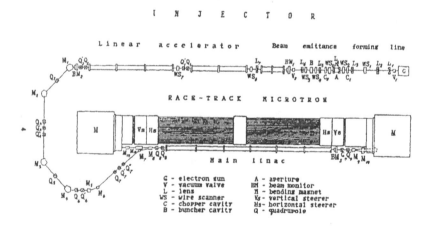

Figure 3.5. Plan view of the MSU CW race-track microtron.

The electron gun beam with energy of 100 keV passes through the chopper-buncher system and then is accelerated in the capture section and preaccelerator up to the energy of 6 MeV. The 6 MeV beam enters the main linac. After the first acceleration up to 12 MeV the beam is reflected and accelerated in the opposite direction up to 18 MeV. Then the beam enters the first orbit of recirculation. This scheme, originally suggested by Kolomensky for a linotron [3.21] and used in the Lund RTM [3.7] and other machines [3.20, 2.19, 3.22], allows sufficient beam clearance from the linac on the first orbit. Beam displacement due to the influence of the injection magnet MS and mirror magnets M9 and M10 is compensated by means of the correct choice of effective lengths and fields of magnets M11 and M12.

The main problem in operating room temperature CW accelerator structures is the high levels of rf power dissipated per structure unit length (from 10 to more than 100 kilowatts per meter). Due to the heating and deformations of these structures: a) there is a frequency shift in the accelerating and coupling cells, b) there are changes in the stopband frequency gap, and c) distortions of accelerating field distributions take place. Actually, the resonant frequency shifts of the coupling cells during the start-up procedure are approximately three times as large as the resonant frequency shifts of the accelerating cells. Because of this the coupling cells tuning frequency are on average 1-1.5 MHz higher than that of the accelerating cells.

As for the rf unit, its important function is to ensure the simple and straightforward start-up of the accelerating sections. The start-up procedure presents one of the most difficult problems for the multi-section linac operation. This is because the resonant frequency shifts that occur during the start-up exceed the resonant curve bandwidth due to heating and deformations of the accelerating structures. A special start-up procedure has been developed, that allows the accelerating sections to be started up separately or simultaneously without resonant frequency tuners. In less than one minute, the linac can be ready for the beam.

Cooling of the accelerator is seen to keep the temperature with a high accuracy. The maximum power dissipated in RTM elements during start-up procedure can reach 780 kW. The nominal power (including rf losses in the accelerator structure walls, as well as the power deposited in the klystron's collectors, consumed by the temperature control system and dissipated in the coils of bending magnets) which must be removed by cooling system is about 630 kW. At the same time the temperature control system must support the temperature of the accelerating structures with the accuracy of 0.05° C.

Particular attention has been given in the project to the bending magnets, as 180° bending magnets comprise an important part of the RTM. The magnetic field induction B_0 in the bending magnets is connected with a synchronous energy gain ΔE:

$$B_0 = \frac{2\pi\Delta E}{\lambda\nu ce}.$$

(3.9)

In this machine for $\lambda = 12.236$ cm and $\nu = 1$ B_0 equals to 1.027 T.

To obtain an energy resolution on the last orbit of several units of 10^{-4} the phase error on the n's orbit $\delta\psi_n$ introduced by field nonuniformity must be less than $1°$. From the relationship

$$\delta\psi_n = \frac{\Delta B}{B} \frac{360 l_n}{\lambda}, \tag{3.10}$$

where l_n is the length of n's orbit, it follows that field uniformity must also be of several units of 10^4.

The next requirement for the bending magnets is connected with the fringing field action on accelerated particles. In order to avoid strong defocusing by the fringing field at low energies, and to reduce the phase errors, reverse field clamps must be used in front of the magnets.

The construction of the bending magnet is shown in Fig. 3.6. It consists of a main yoke, reverse field yokes, poles, separated from the main yoke by homogenizing air gap, the main coils, and reverse field coils. Shims are placed in air gaps between the main yoke and the poles. The configuration and mass of the main yoke, the position of the reverse field yokes, and the thicknesses of air gaps, shims and poles were optimized by numerical calculation. The height of the gap between the poles was chosen to be 60 mm in order to leave enough space for vacuum chamber and correcting plates. The size of the poles is 1500×700 mm^2.

Beam dynamics in the cw RTM with a maximum final energy about 200 MeV differ in some respect from the better studied pulsed RTM. The principal differences stem from the fact that the accelerating gradients in CW machines are an order of magnitude lower. These gradients are limited by the rf power dissipated per unit length, which necessitates, first, a large separation between the bending magnets and, second, 5–15 MeV injection system.

The high average beam current of CW RTMs requires that the particle losses in the acceleration process should be minimized. This minimization is accomplished by properly matching the injected beam emittance and the RTM acceptance. When an output beam with energy spread of the order of 10^{-4} is needed, the requirements for matching the injected beam parameters to the RTM separatrix are even more stringent. All these problems were numerically studied in details.

At present, the injector of this RTM is in operation and the accelerated election beam is used for studying nuclear resonant fluorescence.

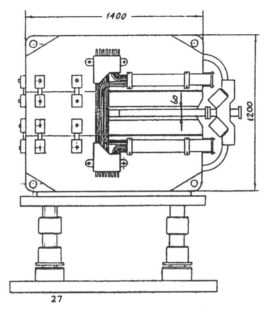

Figure 3.6. Construction of the 180° end magnet.

3.4 Microtron Configurations for GeV Range

The only operating microtron which accelerates electrons to several hundred MeV is the CW race-track microtron MAMI in Mainz (Germany). This machine was considered above and is really a cascade of three RTMs with output energy of the final stage equalling 855 MeV. It is natural investigate whether it is possible to use the microtron philosophy for the energy range up to tens of GeV. This problem was discussed by Herminghaus in [3.23].

The output energy of an RTM is limited to about 1 GeV for two reasons:

1. Magnet weight: for each energy range of an RTM there is an optimum value of flux density in the reversing magnets, which increases with energy. Once the saturation limit is reached, a further increase of energy requires the magnet volume to grow roughly as the third power of energy,

2. Horizontal emittance: its growth per revolution due to synchrotron radiation is given by

$$\Delta \varepsilon = \frac{55\pi^2}{12\sqrt{3}} \frac{r_e \hbar}{mcR^2} \left(\frac{E}{mc^2} \right)^5 \langle H \rangle, \tag{3.11}$$

where

$$\langle H \rangle = \frac{1}{\theta} \int_0^\theta (\gamma \eta^2 + 2\alpha\eta\eta' + \beta\eta'^2) d\psi \qquad (3.12)$$

θ is the deflecting angle of one magnet, α, β and γ are the Twiss-parameters and $\eta = (1 - \cos\psi)$ and $\eta' = \sin\psi$ are the matrix elements for dispersion and angular dispersion within the magnet, respectively. Obviously, there is a rapid increase of $\Delta\varepsilon$ with energy E, such that e.g. in MAMI the horizontal emittance for 99.9% of the beam is increased to 0.14π·mm·mrad. The corresponding beam width is about 5 mm.

It is seen that the energy limit of the RTM scheme is roughly marked by MAMI.

For higher energy, some weight saving theory for the magnets is required and the value of $\langle H \rangle$ has to be reduced. These requirements may be achieved by reducing the deflecting angle per magnet, thus generating 'higher order' microtron schemes (Fig. 3.7). The Bicyclotron, first proposed by Valdner, was closer investigated by Kaiser [3.24], the Hexatron was proposed by Argonne National Lab. [3.25] and the Oktotron, using 8 magnets and 4 linacs, was proposed by Belovintsev *et al.* [3.26], all for the lower GeV range.

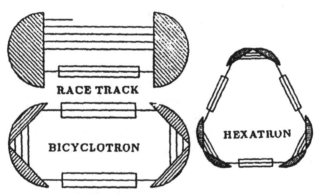

Figure 3.7. Microtron schemes.

The resonance condition mentioned at the beginning closely links the main parameters of such machines. It is obtained from a simple analysis of the orbit geometry as

$$\nu\lambda = \frac{2}{ec} \frac{\Delta E_c}{B} (\theta - \sin\theta), \qquad (3.13)$$

where B is the flux density in the magnets. The energy gain

ΔE_c required per revolution increases very rapidly as θ decreases, as is demonstrated in Table 3.1 using $\lambda = 12$ cm and $B = 1$ Tesla, $N = \pi/\theta$ being the number of magnet pairs.

As a rough estimate, $\langle H \rangle$ is inversely proportional to N^3 [3.27]. Thus, the expression (3.11) is kept constant if N is raised proportional to E, i.e. the number N also represents roughly the output energy limit in GeV. On the other hand, it is seen from Table 3.1 that the energy gain per revolution might even surpass the output energy above 10 GeV, thus spoiling the recirculation concept.

Table 3.1. Microtron parameters

Scheme	θ, deg	N	ΔE_c, MeV
RTM	180	1	5.73
Bicyclotron	90	2	31.5
Hexatron	60	3	99.3
Oktotron	45	4	230
	12	15	11774

This may be solved by allowing more than one magnet pail between adjacent linacs. If N_L is the number of linacs installed, then N/N_L is the number of magnet pairs between adjacent linacs. In that case, Eq. (3.13) is rewritten as

$$\nu\lambda = \frac{2}{ec} \frac{\Delta E_c}{B} \frac{N}{N_L} (\theta - \sin\theta) . \qquad (3.14)$$

Thus, the installed energy gain from Eq. (3.13) is divided by the number of magnet pairs between linacs. This scheme, referred to as 'Polytron', may be used as a recycling accelerator beyond 10 GeV.

Typically, in a polytron the bending angle per magnet is small. Thus, the pole face edges have to be of a stair like shape-to allow adequate beam optics, as shown schematically in Fig. 3.8. Furthermore, the poles are much longer than they are wide. Therefore, it would be advisable to compose each magnet from several, essentially identical iron blocks, each block forming a single stair. This also makes adjustment easier and allows individual trimming of each stair by means of small trim coils. The blocks would be stacked together (leaving a few mm air gap in order to avoid a magnetic short for the trimcoil), and would be excited by one common 'coil' consisting of one single lead.

Figure 3.8. Scheme of a bending magnet. The dotted lines indicate the blocks.

Computations showed that under realistic conditions the beams passing a magnet outside parallel to the pole edge can be shielded very effectively from the fringe field by simple iron tubes and that the field distribution between stair and iron tube gives satisfactory beam optics.

A quadrupole triplet or quintet between the magnets of a pair serves for dispersion matching such that the dispersion vanishes at either end of a pair ('Brown-System').

Detailed examination of the output energy stability and monitoring of longitudinal dispersion showed that it is actually possible to perform acceleration of electrons in RTM for GeV energies.

3.5 A Mobile RTM

The project of a pulsed RTM developed to image concealed narcotics/explosives was described in [3.28, 3.29]. This race-track microtron will produce 40 mA pulses of 70 MeV electrons, have minimal size and weight, and maximal toughress and reliability, so it can be transported on a truck. The carbon/nitrogen cameras for solving this task will be considered later in Part III. The principle RTM parameters are given in Table 3.2.

To increase the reliability, simplify the tuning, and boost the efficiency of mobile RTM design a narrow rectangular asymmetric accelerating structure, rare-earth permanent end magnets, wiggler-like vertical focusing lenses on the return paths, and a beam buncher preceding the linac are used.

A narrow asymmetric rectangular accelerating structure together with short-tail fringe field rare-earth permanent magnet (REPM) end magnets allows the beam to clear the linac after the first acceleration. The beam dispersion after the first end magnet passage is compensated for by the second end magnet. The position of only one magnet need be adjusted to achieve stable phase oscillations.

Table 3.2. RTM parameters

Injection energy	55 keV
Energy gain per turn	5.26 MeV
Number of turns	14
Output energy	10-74 MeV
Output current at 70 MeV	45 mA
Increase in orbit circumference per turn	1 A
Operating frequency	2450 MHz
Klystron power pulsed/average	5 MW/5 (15) kW
End magnet field induction	0.9 T
RTM dimensions	$2 \times 0.6 \times 0.6\,\mathrm{m}^3$

With REPM wiggler-like lenses installed on the return paths instead by end magnet field gradient the beam is vertically focused. These lenses negligibly influence the horizontal motion, no synchronous phase drifts are induced, and the phase stability region is easily maximized. To increase the beam capture and decrease the beam losses beam buncher has been placed at the accelerating structure entrance.

The RTM block-diagram is shown in Fig. 3.9.

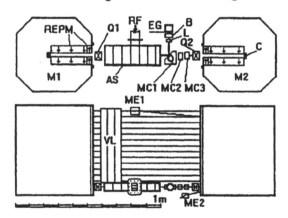

Figure 3.9. RTM schematic: M1 and M2 REPM end magnets, correcting Coils, Accelerating Structure, Electron Gun, Buncher, Quadrupole singlets, MC1-3 chicane magnets, solenoid Lens, Vertical Lenses, and ME1 and ME2 extraction magnets.

The resulting RTM dimensions, $\sim 2.1 \times 0.6 \times 0.5\,\mathrm{m}^3$ and $\sim 2500\,\mathrm{kg}$, are very attractive for use as a transportable source of high-energy γ-radiation.

3.6 Microtron Source for Multicharged Ions

As it has been shown by Shelaev [3.30], the microtron principle of acceleration can be applied to obtain effective production and preacceleration of multicharged ions.

As a rule, the process of multicharged ions production consists of the following steps:

(i) injection of neutral atoms into the ionization region consisting of an electron beam of density j and energy E;

(ii) confinement of injecting particles in the ionization region during the time interval τ;

(iii) extraction of produced ions from ionization region and formation of preaccelerating beam;

(iv) ions separation by their charge and isolation of a beam with necessary charge z subsequent for acceleration.

To produce ions of charge z it is necessary to fulfill two conditions: the electron energy E must be $2-3$ times higher than the ionization potential of the ion of corresponding charge, and the production $jr\sigma_z/e$ has to be of the order of 1, where σ_z is the cross section of ion formation with charge $z+1$. As a rule, the parameters E, j, τ are not controlled, as it takes place in plasma sources, or in controlled conditions the value of τ is so high (as it is in an electron-beam source) that such a source can work only in pulse regime. In all existing sources a portion of ion intensity is inevitably lost after their extraction from the ionization region and subsequent charge separation, as only ions with definite charge are necessary for an accelerator but there are no single charged ions at the output of a source.

In the proposed microtron-like source of multicharged ions two major stages of ion production — confinement of ions in the ionization region and their extraction and separation by charge — are periodically continued in the source itself until a specified atom does not take on the necessary charge. After this an ion beam with a given charge is extracted from the source (in the continuous regime of operation) or can be accumulated in a source and then extracted in the pulsed mode. The pulse duration is determined by the source size and is equal to $10\text{-}15\,\mu s$.

Figure 3.10 illustrates the basic diagram of the proposed source which includes an auxiliary source of single charged ions and two ionization regions which are separated by two charge separation sections. The magnetic structure of the source is practically the same as in a race-track microtron. That is why such a source is named a 'microtron source of multicharged ions' — MSMI.

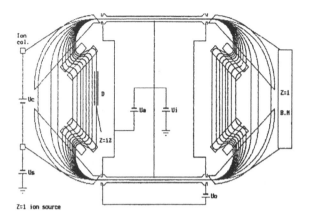

Figure 3.10. Basic design diagram of MSMI.

The auxiliary source provides a beam with a given emittance, and consists of a plasma source of single charged ions, a mass-separator which isolates single charged ions of a given isotope, and a collimator. This auxiliary source is kept under potential $+U_s$ relative to the 'ground', and therefore single charged ions have an energy eU_s at its output.

Both ionization regions are kept under the potential $+U_i$ relative to ground, and thus the energy of injected single charged ions is decreased to the value

$$E_{1i} = e(U_s - U_i) = eU_s(1 - k).$$ (3.15)

Clearly the factor k satisfies the condition

$$k = U_i/U_s < 1.$$ (3.16)

After travelling across the first ionization region the single charged ions are accelerated up to the initial energy eU, and are guided by an additional turning magnet into the second ionization region. As a result of inelastic collisions of single charged ions with electrons the ions with a charge $z \geq 2$ are captured onto closed orbits of the source.

In the ionization region the ions of charge z have an energy E_{1i}, and in the region of separation they are accelerated up to energy

$$E_{sz} = eU_s[1 + (z - l)k],$$ (3.17)

i.e. in MSMI ions are accelerated in a constant electrostatic field through a change of their charge. Therefore the curvature of an ion radius in the region of separation equals

$$R_{sz} = R_{s1}\sqrt{1 + (z - 1)k}/z, \tag{3.18}$$

where R_{sz} is the radius curvature of single charged ions. In the ionization region the following relation between these radii exists: $R_{iz} = R_{i1}/z$. Thus, acceleration of multicharged ions slightly improves their separation in the separation region. To increase the spatial separation of ions with a charge of $z \gg 2$ 4 pairs of plane parallel magnets with a larger field are added. With these magnets the radial separation of ions with charge at any z is sufficient to place additional matching quadrupole lenses on every closed orbit.

As seen from (3.18), the geometry of ion orbits in the separation region is determined by the radius $R_{s1}(M)$ and equals

$$R_{s1} = 45.7\sqrt{AeU_s}/H_s, \tag{3.19}$$

where H_s is the magnetic field intensity (in kOe) in separating magnets, A is the mass number of an ion and eU_s in keV. If the value of k is chosen constant for any A the same z-orbits for ions with different A can be allowed by the proper selection of $H = H(A)$.

Extraction of ions from the source is carried out by the electrostatic deflector D which is placed at the last closed orbit. The deflection can be performed in continuous or pulsed regime depending on regime of supply. There are two identical three-electrode guns in each ionization region. These guns form dense electron beams with velocities which have opposite direction to the velocity of ion beams. Thus the longitudinal velocity of circulated ions is stabilized due to multiple elastic collisions with electrons. The stability of the transverse motion of ions with a charge $z \geq 2$ is provided by the proper choice of the of magnetic system parameters.

The estimates, which were performed by Shelaev, show that, for instance, for helium-like ions $(z = N - 2)$ their intensity from MSMI will be of the order of $10^{11} \div 10^{12}\,\mathrm{s}^{-1}$ and such ion current is sufficient for modern nuclear physics.

Chapter 4

Free Electron Laser

The laser has revolutionized optics. Many other fields of science and technology, previously unrelated, have benefited from the use of coherent radiation in the infrared, visible, and ultraviolet regions of the spectrum. The development of lasers that can generate radiation at still shorter wavelengths, comparable with atomic dimensions, can likewise be expected to lead to a vast array of new scientific and technological developments. The electromagnetic spectrum is shown in Fig. 4.1, where currently available sources of radiation are also given. One can see that sources are lacking both in the infrared (IR) region ($1\,\mu\mathrm{m} \leq \lambda \leq 1\,\mu\mathrm{m}$) and in the ultraviolet (UV) and shorter wavelength region ($\lambda \leq 1000\,\text{Å}$). Free-electron lasers (FEL) and X-ray lasers could presumably fill these gaps.

Many of the most promising uses of the FEL can be envisaged in those wave-length regions where efficient tunable high-power lasers do not operate. Figure 4.1 suggests that there would be particular interest in FELs operating in the infrared or far-infrared region where they are expected to furnish a considerable amount of power compared to conventional laser sources. This region is of primary interest for solid-state physics, and at shorter wavelengths FELs may be an interesting source for chemical and molecular-spectroscopy studies.

In this chapter we consider the physical basics of FELs — new laser-like sources of radiation, in which electron accelerators, in particular microtrons, are main components.

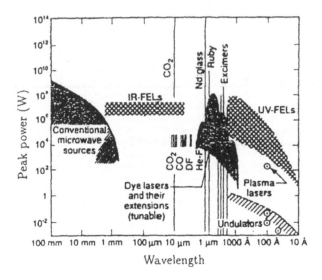

Figure 4.1. The power and range of some sources of radiation: conventional microwave sources (tubes, klystrons, gyrotrons); lasers; plasma lasers (indicated by ⊙); undulators on third-generation synchrotron radiation facilities. Possible FEL performance is indicated by the cross-hatched regions (from [4.1]).

4.1 Physical Basics of FEL

A relativistic beam of electrons could be emitted by passing the beam through a periodic electric or magnetic field. The origin of such spontaneous emission is simple to understand. In the 30's Weizsäcker [4.2] and Williams [4.3] have shown that the bremsstrahlung of very fast electrons was relatively easy to calculate in the inertial reference frame where the electrons are at rest. In this frame, the electrostatic potential of the passing nucleus is 'seen' as high-frequency light with a wide spectrum. This light induces Compton scattering of the electrons 'at rest'. According to this approach, any structure, artificial or natural, in which the electric or magnetic field is a periodic function of one coordinate will be seen as a nearly monochromatic and quasi-plane wave by a travelling fast electron. The relative line-width of this wave would be equal to the inverse number of spatial periods of the static field (static in the laboratory frame).

The two experiments [4.4, 4.5] performed in Stanford University in 1975 and 1977 were a turning point in the research on stimulated electromagnetic radiation by relativistic electrons. The results demonstrated the possibility that this mechanism can be used in the development of a new class of tunable high-power lasers, the so-called free electron laser (FEL) since the generation of coherent radiation is produced by means of free electrons.

The existing modern FELs cover a wide range of the electro-magnetic spectrum, their feasibility has indeed been proved from the visible to the millimeter region. Furthermore, existing projects tend to improve the working range down to the vacuum ultraviolet and even to the X-ray region.

The principle of operation of an FEL is clearly seen from Fig. 4.2 which presents canonical scheme with two mirrors, forming a Fabri-Pérot resonator. The element transforming the longitu-dinal non-radiating electron motion into transverse motion is the undulator, i.e., a magnetic device with alternating N-S poles. The mechanism of spontaneous emission in a FEL is purely classical and is due to the acceleration of the electron during its motion in the undulator.

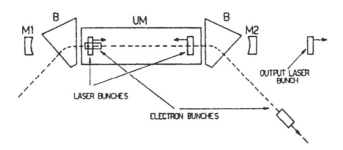

Figure 4.2. FEL oscillator operating with a bunched e-beam. In this con-figuration the photon round trip period is twice the time (distance) between two adjacent electron bunches, so that two photon bunches are running back and forth into the optical cavity (B – bending magnet, M1 – fully reflecting mirror, M2 – output coupler mirror).

Since the spontaneous emission in undulator magnets is a prerequisite for the FEL action itself, we will first analyze the spectral feature of the spontaneously radiated power and will use for this aim a simple approach, developed in [4.6], just to understand the spectrum shape and the selection mechanism of the central emission frequency.

Let us consider an electron moving in a magnetic field with components

$$\vec{B} \equiv \left[0, B_0 \cos\left(\frac{2\pi z}{\lambda_u}\right), 0\right], \qquad (4.1)$$

where λ_u is the wavelength of the undulator. The Lorentz force induces a transverse velocity given by

$$v_t = -\frac{cK\sqrt{2}}{\gamma} \sin\left(\frac{2\pi z}{\lambda_u}\right), \qquad (4.2)$$

B_0 is axial magnetic field, and the dimensionless parameter

$$K = \frac{eB_0\lambda_u}{2\sqrt{2}\pi mc^2} \qquad (4.3)$$

is usually called the undulator strength. It determines the magnitude of electron oscillations in the undulator. The average longitudinal velocity will therefore be given by

$$v_0^2 = v^2 - \langle v_t^2 \rangle = c^2 \left[1 - \frac{1}{\gamma^2}(1 + K^2)\right], \qquad (4.4)$$

where $\langle\rangle$ denotes an average over the undulator period.

In the transverse plane the electron is oscillating at a frequency

$$\omega_u = \frac{2\pi c}{\lambda_u}. \qquad (4.5)$$

In the electron rest frame, i.e., in the frame in which the average longitudinal velocity is zero, Eq. (4.5) transforms into

$$\omega' = \frac{\omega_u}{\sqrt{1 - (v_0/c)^2}}. \qquad (4.6)$$

In this frame the electron radiates isotropically just at the frequency ω'.

Going back to the laboratory frame, the frequency ω' undergoes a Doppler shift so that the central emission frequency reads:

$$\omega_0 = \sqrt{\frac{1 + v_0/c}{1 - v_0/c}} \cdot \omega' = \frac{4\pi c}{\lambda_u}\frac{\gamma^2}{1 + K^2}, \qquad (4.7)$$

or can also be writtenwhat as:

$$\lambda_0 = \frac{\lambda_u}{2\gamma^2}(1 + K^2). \qquad (4.8)$$

The width of the emitted line can be inferred from the duration Δt of the light pulse. The interval of time can immediately be obtained from

$$\Delta t = \frac{L_u}{v_0} - \frac{L_u}{c} \simeq \left(1 - \frac{v_0}{c}\right)\frac{N\lambda_u}{c} \simeq \frac{2\pi N}{\omega_0}, \qquad (4.9)$$

where L_u and N are the length and number of periods of the undulator respectively $(L_u = N\lambda_u)$.

Finally, from the uncertainty relation, it follows that

$$\Delta\omega\Delta t \simeq \pi, \tag{4.10}$$

and thus

$$\left(\frac{\Delta\omega}{\omega_0}\right) = \frac{1}{2N}. \tag{4.11}$$

The relative bandwidth (4.11) will be referred to from now on as the *homogeneous* bandwidth. For typical undulators we have $N = 50$ and thus a homogeneous bandwidth of 1%. Such a small width is the distinctive feature of the undulator radiation as compared to that from synchrotron emission which is considerably wider owing to a very short light pulse. The undulator spectrum shape is given by

$$\frac{dP}{d\nu} \propto \left(\frac{\sin \nu/2}{\nu/2}\right)^2, \quad \nu = 2\pi N \frac{\omega_0 - \omega}{\omega_0}. \tag{4.12}$$

As a final comment let us note that expression (4.12) gives only partial information; in fact it does not contain anything about, for example, the angular distribution or the harmonic content.

The frequency and wavelength of the radiation emitted at an angle θ with the undulator axis are

$$\omega_0 = \frac{4\pi c\gamma^2}{\lambda_u} \frac{1}{(1 + K^2 + \gamma^2\theta^2)}, \quad \lambda_0 = \frac{\lambda_u}{2\gamma^2}(1 + K^2 + \gamma^2\theta^2). \tag{4.13}$$

We will call the emission process assisted by a radiation field copropagating with the electrons, *stimulated emission* . In this case a variation of intensity of the input field, proportional to the electron energy variation, can be observed.

We consider a wave with a transverse electric field

$$E_l = E_0 \cos(\omega t - kz + \phi_l), \tag{4.14}$$

copropagating with the electrons in the longitudinal direction z; ϕ_l is the relative phase between field and electron. The field (4.14) couples with the electron transverse motion and induces an energy variation governed by the equation

$$\frac{d\gamma}{dt} = \frac{eE_l v_t}{mc^2}, \tag{4.15}$$

which once coupled with Eq.(4.2), yields

$$\frac{d\gamma}{dt} = -\frac{eE_0K}{mc\gamma\sqrt{2}}\sin\psi, \qquad (4.16)$$

with ψ defined as

$$\psi = \left(k + \frac{2\pi}{\lambda_u}\right)z - \omega t - \phi_1. \qquad (4.17)$$

From Eq.(4.2) we also get $(\dot{z} = v_0)$:

$$\frac{d^2z}{dt^2} = c\frac{\dot{\gamma}}{\gamma^3}(1 + K^2), \qquad (4.18)$$

which once inserted into Eqs.(4.16) and (4.17) gives the equation governing the evolution of the coupled electron-radiation system, i.e.,

$$\frac{d^2\psi}{dt^2} = -\Omega^2\sin\psi, \qquad \Omega^2 = \frac{e^2E_0B_0}{(m_0c\gamma)^2}. \qquad (4.19)$$

Equation (4.19) is particularly impressive; it states indeed that FEL dynamics can be modelled using a simple pendulum-like equation. The above equation can be solved exactly by means of elliptic Jacobi functions. It is, however, more convenient to avoid an unpleasant mathematical method and treat Eqs.(4.19) in two distinct regimes.

$$I) \quad \textit{Small-signal regime} \quad \Omega^2 \ll (c/L_u)^2. \qquad (4.20)$$

$$II) \quad \textit{Strong-signal regime} \quad \Omega^2 \approx (c/L_u)^2. \qquad (4.21)$$

The first case corresponds to weakly perturbed electron motion, the second to strong electron-field coupling. The essential features of these regimes are reported in the phase-space plot of Fig. 4.3. The region (I) corresponds to the open orbits above the separatrix, the region (II) to the closed orbits below the separatrix. In region (I) we have an effective energy exchange between electron and field, while in region (II) saturation effects arise.

Since, according to the energy conservation, an energy loss of the electron will reflect as an increment of the input field intensity, we will define the FEL gain as

$$G = -mc^2\frac{\Delta\gamma}{W_0}, \quad W_0 = \frac{1}{8\pi}E_0^2V, \quad V \equiv \text{mode volume}. \qquad (4.22)$$

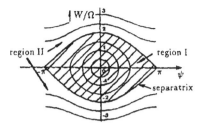

Figure 4.3. FEL pendulum-like phase space. In region I (dashed region inside the separatrix) the motion is periodic (saturation). In region II the phase-space trajectories are open (small-signal regime).

The quantity $\Delta\gamma$ can easily be calculated in the region (I), where we can define naturally the following perturbation parameter;

$$\Omega_R^2 = \left(\Omega\frac{L_u}{c}\right)^2 \ll 1 . \tag{4.23}$$

Thus, we can use perturbative methods, and averaging over the phase ϕ_1 we get after some rather straightforward mathematical calculations the following simple expression for the gain G:

$$G = -g_0\pi\frac{d}{dv}\left(\frac{\sin \nu/2}{\nu/2}\right)^2 , \tag{4.24}$$

where g_0 is the gain coefficient determined by the e-beam and optical cavity parameters. Equation (4.24) states the important result that the gain profile is proportional to the derivative of the spontaneous spectrum (see Fig. 4.4).

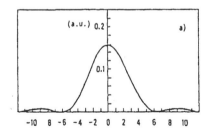
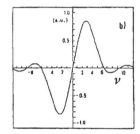

Figure 4.4. Spontaneous emission (a) and gain (b) curves.

An e-beam produced by an rf accelerator will exhibit a longitudinal structure of the type displayed in Fig. 4.5 namely a macropulse of the order of tens of microseconds and a microstructure of the order of tens of picoseconds, fixed by the phase-stable angle; furthermore the distance between successive bunches is fixed by the period of the rf field.

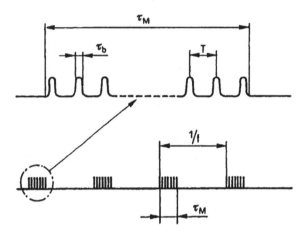

Figure 4.5. Time structure of an rf electron beam: τ_M – macropulse duration, T – bunch-bunch distance, τ_b – micropulse duration, f – repetition frequency.

4.2 Electron Accelerators for FELs

Practically all the types of electron accelerators — synchrotrons, storage rings, electrostatic accelerators, microwave and induction linacs, microtrons of the conventional and race-track types — are used in FEL projects. The FEL oscillatory regime has been obtained for many wavelengths — from ultraviolet to millimeter-band range. It should be noted that many types of gas and solid-state lasers of conventional type have also been developed in the visible and infrared regions. Their characteristics including high and ultrahigh power, tunability, possibilities to obtain short and ultrashort light pulses, etc., are sufficient for the most applications. Recently of the greatest practical interest are the ultraviolet (UV), vacuum ultraviolet (VUV), infrared and especially far-infrared (FIR) regions for FEL applications.

Progress in the FEL development depends to a great extent on the characteristics of electron beams formed by accelerators. The choice of the type of accelerator depends on the particular characteristics of the FEL under development, but as a rule, the requirements for beam quality (emittance, energy spread) and beam current are very rigid. In most cases the accelerators have to be specially developed to obtain the beam quality necessary for FEL operation. For instance, the development of pre-buncher systems and photocathodes in linacs is responsible for the high efficiency of the FELs in the visible and infrared regions.

The development of FELs for the UV (and shorter) region is very promising. As opposed to the active media usually used in lasers, the electron beam in FELs has no principal frequency restrictions. However, the progress in development of the UV (and the shorter wavelength range) FELs is not impressive because of strict limitations imposed on the beam quality and because of the lack of highly-reflective materials that could be used for mirrors in these regions. In practice, only storage rings of several hundred MeV energy and with good beam quality can be considered as promising sources for the UV, VUV and X-ray FELs.

The infrared FEL projects can be divided into two types. The FELs of the first type use the quasicontinuous electron beam generated by several MeV electrostatic accelerators. One FEL based on a 3 MeV electrostatic accelerator is capable of delivering multikilowatt-level microsecond pulses of coherent radiation in the wavelength range 390–1000 μm.

Pulsed accelerators — linacs and microtrons with microwave power supplies — are used in the FELs of the second type. The electron beam in these accelerators is a sequence of short bunches (1 mm-1 cm long) separated by the distance which is a multiple of the wavelength of the microwave power supply. The pulse duration can reach several tens of microseconds.

The FEL radiation in such devices is also a sequence of powerful light micropulses with the same time structure following that of the electron bunches. These pulsed FELs should be considered as very promising sources of powerful picosecond pulsed radiation, tunable in a wide spectral range. For general applications in the infrared and far-IR FEL the pulsed microtron provides an adequate beam well matched to the FEL [4.7, 4.8].

The emittance ε_N of the microtron electron beam at the Nth orbit can be estimated to be

$$\varepsilon_N \leq \frac{\lambda}{4000N}, \qquad (4.25)$$

where λ is the wavelength of the accelerating cavity. It is this emittance that has to be matched to that of the FEL magnet system described by a similar expression, where N would be the number of periods of the undulator and in place of the wavelength of its periodic structure.

The typical values of the energy spread and emittance of an electron beam accelerated in a microtron are $\Delta\gamma = 0.005$ and $\varepsilon = 5 \times 10^{-4}$. Introducing these into the expression of the starting current I_s of an FEL we may calculate the threshold power for FEL operation at different wavelengths and the result is shown in Fig. 4.6.

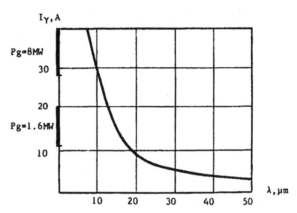

Figure 4.6. The dependence on the wavelength of the threshold power of an FEL with a microtron (from [4.8]).

We consider below microtron based FELs in operation as well as interesting new projects.

4.3 Cherenkov FEL

Slow-wave structures, which achieve velocity synchronism between the propagating wave and the interacting electron beam, are considered as potential candidates for the generation of power in the millimeter and submillimeter regions of the spectrum, since they overcome many of the problems related to the excitation of microwave cavities with dimensions of the order of the wavelength.

A number of experiments exploiting the emission of Cherenkov radiation in dielectric-loaded waveguides have been performed in the past, mainly in the millimeter region utilizing electron beams at energy $\leq 1\,\mathrm{MeV}$. In this type of device a beam of relativistic electrons passes at grazing incidence above the surface of a dielectric-loaded waveguide, exciting TM-like surface waves. The longitudinal component of the evanescent electric field causes a bunching of the electrons, which give rise to stimulated emission. In the synchronism condition for a single-slab geometry the radiation is emitted at a wavelength given by

$$\lambda = 2\pi d\gamma(\varepsilon - 1)/\varepsilon \qquad (4.26)$$

in the approximation $\gamma \gg 1$, $\lambda \gg d$, where d is the film thickness, ε its dielectric constant, and γ the relativistic factor of the electrons. The coupling efficiency C between a single electron and the TM wave is expressed by exponential behaviour

$$C = \exp(-2\pi x/\lambda\beta\gamma), \qquad (4.27)$$

where x is the distance of the electron from the film surface. Operation at high energy is therefore interesting since it can allow either an increase in the coupling efficiency for a given wavelength or achieving a shorter wavelength for a given coupling efficiency.

An experiment utilizing a radio-frequency-driven 5-MeV microtron as the electron-beam source has been designed and performed at the ENEA sub-millimeter free-electron-laser facility in Frascati [4.9]. The layout of the experiment is sketched in Fig. 4.7.

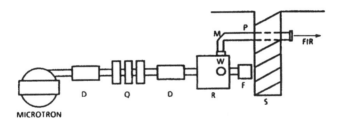

Figure 4.7. Layout of the experiment. D — deflecting coils, Q — quadrupoles, R — resonator chamber, W — quartz window, F — Faraday cup, M — plane mirror, P — sealed light pipe, S — concrete shield.

A quasioptical resonator provides feedback and confinement of the radiation (see Fig. 4.8). Two pairs of permanent magnets, placed symmetrically sideways at the entrance to the resonator, allow the electron beam to be injected above the input mirror and then be displaced so as to travel parallel and close to the surface of the dielectric film. The gap between the magnets can be adjusted in order to optimize the coupling of the electrons to the dielectric film. In the interaction region the electron beam is focused by a triplet of quadrupoles to a small elliptic cross section $(6 \times 3\,\text{mm}^2)$.

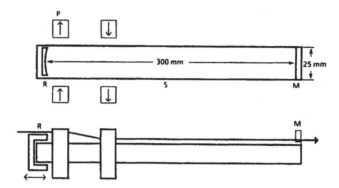

Figure 4.8. Quasioptical resonator. S — diamond machined copper substrate, R — cylindrical mirror (45 cm radius of curvature), P — permanent magnet SmCo, M — grid reflector output coupler 500 lines/inch.

Between the resonator output and the z-cut quartz window of the vacuum chamber, a copper horn was inserted (see Fig. 4.9) which collects only the radiation coming out of a narrow slit $(15 \times 3\,mm^2)$ at the end of the film surface. The horn removes the microwave background associated with the propagation of an rf-modulated electron beam through the metal structure of the transport channel and also launches the emitted surface wave into the light pipe, avoiding the strong diffraction which would be present with free-space propagation.

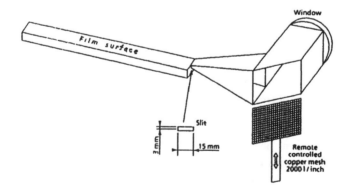

Figure 4.9. A view of the dielectric-loaded waveguide and of the copper horn utilized to collect the millimeter and submillimeter radiation.

In the experimental configuration two different polyethylene films, with thicknesses of 25 and 50 μm, respectively, have been successfully operated. The wavelength of the emitted radiation is 1660 µm for the 50-µm-thick film and 900 µm for the 25-µm-thick film.

In Fig. 4.10 a typical signal detected by the InSb bolometer is shown with the electron-beam current pulse. An interesting feature is the signal growth during the macropulse which can be ascribed both to nonlinear behaviour with respect to the variation of the electron current and to feedback in the resonator structure. Analysis of the signal amplitude as a function of the distance of the electron beam from the dielectric film surface gave a clear indication of a Cherenkov emission process. The maximum signal was obtained not for the maximum transported current, but for the maximum current density above the film surface, which corresponds to a position of the beam centroid approximately on the surface of the dielectric film. A displacement of the beam centroid of 1.5 mm from the film surface caused a decrease of the output signal of a factor of 4, just as one would expect from the coupling coefficient C integrated over the electron-beam transverse distribution. Analysis of the emitted radiation has been performed and 80% polarization for the electric field perpendicular to the film surface was found, in agreement with the excitation of TM-like surface waves.

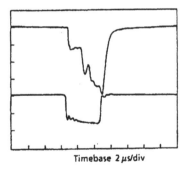

Timebase 2 µs/div

Figure 4.10. 50-µm-film operation. Upper trace: InSb-detector signal (200 mV/div). Lower trace: electron-beam current (100 mA/div).

A measurement of the power emitted with the 50-µm film was performed and it was shown that power up to 50 W in pulses of 4-µs duration can be generated at wavelengths of 1.6 and 0.9 mm (with two different dielectric loaded waveguides).

The measurements performed on the output radiation of the Cherenkov FEL relevant to the spectral content, polarization, and sensitivity to the electron-beam steering are in close agreement with the expected characteristics. The output power is about 6 orders of magnitude higher than would be expected from spontaneous emission. The observed behaviour of the detected signal, as a function of the electron current amplitude, strongly suggests the occurrence of coherent emission from the current modulation at the resonant frequency, due to the bunched nature of the electron beam.

4.4 FEL in the Millimeter-Wave Region

Several issues arise in a long wavelength FEL driven by an rf accelerator as a result of the short electron bunch duration. When the bunch length is comparable to the operating wavelength λ, a considerable amount of is expected even with no feedback in the resonator. On the other hand, the buildup of the FEL pulse is strongly affected by the 'slippage', i.e. the lack of overlapping between the wave packet and the electron bunch due to their different velocities. In free space the slippage length at the end of the interaction region is $\delta = N\lambda$, where N is the number of undulator periods; in the millimeter-wave region it can be quite large compared to the electron bunch length and may preclude the onset of oscillation. In a waveguide it is possible to control the group velocity of the excited mode by a proper choice of the waveguide transverse dimension and therefore reduce the slippage. A particular condition is found when $\lambda_u \simeq 2\gamma_z b(1 - 0.8/2N)$ for the fundamental mode in a planar waveguide, where λ_u is the undulator period, γ_z is the relativistic factor of the electron associated with the drift motion along the undulator with velocity β_z, and b is the waveguide gap. In this condition a broad band gain curve with a relative bandwidth $\Delta\lambda/\lambda \simeq 1/\sqrt{N}$ is obtained around a central frequency having a group velocity equal to the electron velocity; it is therefore called the 'zero-slippage' condition.

To test the features of the waveguide FEL, and in particular the operation at zero slippage, an experiment was designed and performed at the ENEA submillimeter FEL facility in Frascati [4.10]. The 5 MeV microtron delivers short electron bunches with 15 ps duration spaced at the frequency of 3 GHz. The bunches form a train (macropulse) with duration up to 4 μs, which is repeated at a maximum frequency of 40 Hz. The peak current in the bunch is $I_p = 6$ A.

The undulator and resonator assembly is shown in Fig. 4.11. The compactness of this waveguide FEL has been obtained by using a linearly polarized permanent magnet undulator with a period $\lambda_u = 2.5$ cm and a length of 22.5 cm. It provides a peak magnetic field of 6.1 kG on an axis at a 0.63-cm gap between its poles, corresponding to a value of the undulator parameter $K = 1$.

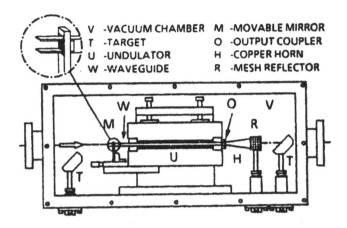

V -VACUUM CHAMBER   M -MOVABLE MIRROR
T -TARGET   O -OUTPUT COUPLER
U -UNDULATOR   H -COPPER HORN
W -WAVEGUIDE   R -MESH REFLECTOR

Figure 4.11. A view of the undulator and resonator assembly.

The two assemblies of the undulator poles are directly clamped on a straight section of WR42 rectangular copper waveguide which has a length of about 30 cm and internal dimensions $a = 1.067$ cm and $b=0.432$ cm.

Spontaneous emission measurements were carried out without mirrors in the resonator. About 20 W of power was generated in a composite broad-band emission covering the spectral range from 1.5 mm to 4 mm, showing the coherent excitation of discrete frequencies which are harmonics of the 3 GHz radio frequency due to the bunched structure of the e beam.

The first oscillation of the laser was observed in a narrow band at a wavelength of 2.43 mm (123 GHz) with a fixed resonator length of 29.34 cm and a 2% output coupler. The spectral characteristics of the emitted radiation were analyzed by means of the FPI and are shown in Fig. 4.12.

The measured power level and the narrow line width observed are clearly explained by the growth and saturation mechanism for a coherent signal at a single harmonic of the 3 GHz, which is stored in the resonator and gains energy from the electron bunches that synchronously enter the undulator. The narrow linewidth of the emitted radiation indicates that many bunches contribute to the buildup of the oscillation, which occurs over a time duration

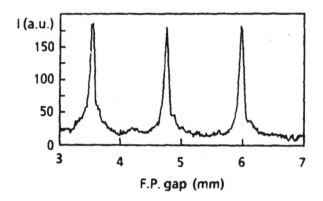

Figure 4.12. Fabry-Pérot interferogram of the FEL output at L=29.34 cm.

much longer than the electron bunch and the bunch spacing itself. When the radiation is generated inside a cavity, the harmonic components of the electron beam current modulation have to drive resonant modes of the cavity within the FEL resonance band.

Absolute power measurements have been carried out at the output of the light pipe by using an optoacoustic microwave energy meter and have been compared to the pyroelectric detector measurements. The dependence of the emitted power on the amount of output coupling has also been investigated. A maximum power of about 300 W was obtained with 6% output coupling.

4.5 Far-Infrared FEL

Laser output in the far-infrared range from a 20 MeV microtron based free electron laser was obtained in the early 90's in the Bell Laboratory by Shaw and Chichester [4.11].

A schematic diagram of the FEL apparatus is shown in Fig. 4.13. The electron beam is focused at the entrance of the electromagnet helical undulator by two quadrupole doublets. Two magnetic dipoles located between the quadrupole lens translate the electron beam onto the undulator axis, which is colinear with the optical axis formed by two 10 m radius of curvature copper mirrors spaced 15 m apart. Vertical and horizontal deflection coils located at the entrance of the undulator position the electron beam on the undulator axis and provide a small angular deflection to minimize the betatron amplitude of the electron trajectory in the undulator. The electron beam is deflected at the exit of the undulator into a Faraday cup by a vertical deflection coil and detected. Computer tuning of the FEL is accomplished by simultaneously adjusting two horizontal deflection coils at the entrance of the undulator, a vertical deflection coil at the exit, and current to the undulator to maintain the electron beam close to the axis and minimize outgassing near the exit Faraday cup. The undulator is magnetically shorted at both ends to further facilitate steering the beam close to the axis. The K parameter for the undulator is approximately 1 for a helical field slightly greater than $500\,\text{G}$ and a $20\,\text{cm}$ period.

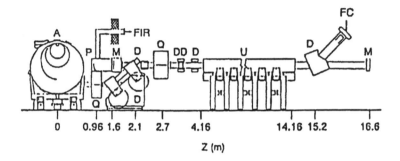

Figure 4.13. Microtron accelerator, A. The beamline elements are as follows: Q — quadrupole doublet, D — dipole bending magnet, U — helical undulator magnet, FC — Faraday cup beam dump. The optical cavity mirrors are M and P is parabolic mirror that focuses FIR into $10\,\text{m}$ transport line.

The microtron accelerator produces $16\,\mu s$ macropulses with peak current of $90\,\text{mA}$ at energies of 19-$20\,\text{MeV}$ and $30\,\text{Hz}$ repetition rate which corresponds to micropulse space separation of $10\,\text{cm}$. The temporal profile of the electron beam macropulse measured at the exit Faraday cup is shown in Fig. 4.14 as well as FIR time structure.

The FIR output wavelength dependence on undulator current shown in Fig. 4.15 is determined utilizing a Fabry-Pérot interferometer constructed with metal mesh.

The solid line slope of $0.3\,\mu\text{m/A}$ is fitted with the results of magnetic field measurements with undulator current using the relation

$$\lambda = \frac{\lambda_u(1+K^2)}{2\gamma^2}. \tag{4.28}$$

The output power dependence on wavelength, corrected for the

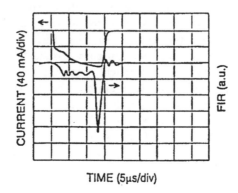

Figure 4.14. FIR time structure measured at 220 μm. Upper trace: electron beam current. Lower trace: Ge:Ga detector signal.

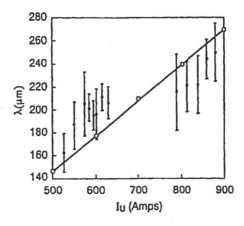

Figure 4.15. Wavelength of FIR output versus undulator current.

wavelength dependence of the detector, is given in Fig. 4.16.

The electron beam micropulse length is about 6 mm, which is less than the optical pulse slippage parameter $(N\lambda)$ which is 14 mm at 280 μm. Here N is 50, the number of undulator periods, and the slippage parameter gives the distance that a photon gains on the electron in the time the electron traverses the undulator. In FELs designed to have one micropulse present in the optical cavity at a time, the round trip gain is decreased when the micropulse length is appreciably less than the slippage parameter, resulting in more difficult operation at long wavelengths. The increased power output with wavelength is evidence for micropulse

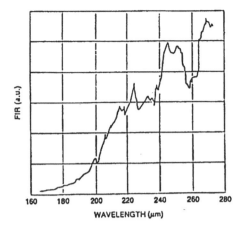

Figure 4.16. FIR output power versus wavelength.

radiation interference. The cavity peak power is estimated to be about 100 W.

Thus, rapid and broad wavelength tuning of a microtron based FEL have been demonstrated at FIR wavelengths from 160 to 280 μm with a frequency resolution of about 0.5 cm^{-1}.

The Korean Atomic Energy Research Institute and the Budker INP (Novosibirsk) developed a design and manufactured the main parts of a compact free ectron laser with emission power of one watt in the infrared region of 25 to 30 microns on the basis of 10-cm microtron with the magnetron power of about 1.5 MW and pulse duration 6.5 μs [4.12]. The output electron energy is 7.5–8 MeV.

The FEL undulator is an equipotential-bus undulator [4.13]. Its upper part consists of straight buses, that is from the top to the left and from the top to the right alternately clenched by poles. The plates of permanent magnets are inserted between the poles. The lower part of the undulator is inversion symmetry to the upper one. The undulator period equals 1.25 cm, number of period equals 160, the gap is 0.5 cm, and K parameter equals 0.4–0.6. It seen, that a rather long undulator with a small period is achieved. High quality of an electron beam from the microtron allows to transmit ~80% of the current through the relatively narrow and long undulator gap. In spite of not so high electron current (~ 25 mA), rather large gain gives an opportunity of reaching a lasing in this device.

4.6 Unconventional Relativistic Oscillators Driven by the Microtron

New concepts of coherent light generating devices using relativistic electron beams have been recently considered.

4.6.1 Reflex Free Electron Laser

In the classical FEL the space bunching arises from velocity modulation of the beam caused by interaction with an electromagnetic wave. In 1977 a new scheme, namely, an optical klystron composed of two undulators separated by a dispersive section, was proposed by Vinokurov and Scrinsky [4.14]. In such a device the space bunching in the dispersive section arises from the energy modulation after the interaction of the beam with the wave in the first undulator. For ultrarelativistic beams this reduces the starting current of the FEL.

Kapitza and Semenov [4.15] considered an FEL scheme analogous to the optical klystron but with simultaneous bunching and reflection of the electron beam by the dispersive magnet system.

In the general case, the longitudinal bunching appearing during the passage of the beam through the dispersive system is disturbed by the transverse motion. If the transformation matrix of the dispersive system in basis $(x, \dot{x}, \Delta\gamma/\gamma, \Delta S)$ (where $x, \dot{x}$ is the transverse displacement with respect to a central ray and its derivative, respectively, γ is the electron energy divided by its rest energy, and ΔS is the path length difference from that of central ray),

$$\begin{bmatrix} M_{11} & M_{12} & M_{13} & 0 \\ M_{21} & M_{22} & M_{23} & 0 \\ 0 & 0 & 1 & 0 \\ M_{41} & M_{42} & M_{43} & 1 \end{bmatrix} \quad (4.29)$$

has elements M_{41}, M_{42} equal to zero, then the bunching at the system's exit is due only to the energy modulation of the beam at the system's entrance. From the general relations for arbitrary beam-handling magnet systems it follows that the condition $M_{41} = M_{42} = 0$ is equivalent to $M_{13} = M_{23} = 0$. If this condition is satisfied the system is called achromatic. Then the longitudinal dispersion of magnet system is determined only by the matrix element M_{43}.

In experiments it is desirable to have the possibility to vary the dispersion of the magnetic system from zero to a maximum limited by the energy spread. But in the process of dispersion variation, the system should still remain achromatic in order to avoid the influence of beam transverse dimensions and angular spread on longitudinal bunching. These requirements are satisfied by an achromatic reflecting magnet. The general layout of the reflex FEL proposed is schematically shown in Fig. 4.17.

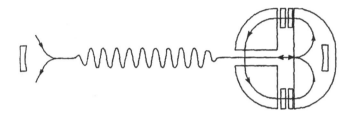

Figure 4.17. Reflex FEL layout.

The reflecting magnet contains two sector magnets (π-sector and $\pi/2$-sector) separated by a quadrupole lens section at each side of the symmetry plane. At the first half-turn after the entrance into the reflecting magnet the electrons deviate from the central ray according to the magnitude of the beam energy modulation, and leave the first sector (π-sector) as a parallel beam. Then the beam should be refocused by a system of quadrupoles placed between the sectors to compensate for the path length difference before electrons cross the symmetry plane.

The reflex FEL combines the properties of the optical klystron, in particular, its smaller starting current (in comparison with the classical FEL), and the possibility to reduce the optical cavity length as well as the whole laser length. The reduction of optical cavity length brings about the reduction of the norms of the cavity eigenfunctions, and, as a consequence, a larger gain. The decrease of radiation rise time is another advantage. This is especially important for lasers based on pulse accelerators.

4.6.2 Interaction of Electrons with Open Resonators Located Inside a Microtron

The electrons inside the microtron chamber move in circular orbits, and thus emit radiation. It is of interest to work out conditions under which the spontaneous synchrotron emission of the electrons could be built up into coherent synchrotron radiation. In this case open resonators placed in the microtron chamber can be considered for exciting synchrotron radiation.

For instance, an open resonator for generating induced synchrotron radiation over a substantial arc of the orbit is proposed in [4.16]. Theoretical investigation of radiation produced by relativistic electrons moving in an open resonator in the whispering-gallery mode field was considered in [4.17].

The peculiarity of the whispering-gallery mode is that the field of this mode is concentrated near the resonator wall. In the first of two systems a section of a barrel shaped open resonator is used. The whispering-gallery standing wave with an azimuthal dependence can be excited inside such a resonator. Electrons move near the inner caustic surface of the wave. Formulas obtained show that this system can develop considerable gain.

The dependence of gain on the electron energy is not so sharp as in usual FELs. This is because of the more efficient interaction of the electrons travelling in the circular orbit with the azimuthal component of the electric field of the wave and additionally by a more efficient 'gyrotron' bunching in a magnetic field. However, as in an optical klystron, the electron energy spread restricts the gain of this system. Besides, as the whispering-gallery wave has a sharp radial and azimuthal field distribution, the gain very sharply depends on the transverse dimensions of the beam.

Kapitza and Kleev [4.18] considered the system with a three-mirror open resonator. An electron moving in circular orbit in the microtron successively interacts with two light beams in the resonator. The first interaction changes electron energy and its transverse momentum. A section of the electron trajectory between the interaction regions is used as a buncher. In the second interaction the electron can transfer a portion of its energy to the electromagnetic wave.

However, in the system under consideration the mechanism of electron-wave interaction is more complicated than in the optical klystron. Two-dimensional motion of the electrons in very nonuniform fields of the resonator was found only by computer simulations.

In both the resonator systems considered the optimal wavelengths belong to the short millimeter band. Although the latter system has a lower gain, one should keep in mind that the power losses in the three-mirror resonator are considerably lower than those in the barrel-shaped one.

4.7 High Power Infrared FEL Driven by a Microtron-Recuperator

During recent years a very promising project for a high-power FEL using a race-track continuous microtron-recuperator has been under development in Novosibirsk Scientific Centre [4.19]. The goal of this project is the creation of a tunable infrared FEL ($\lambda = 1 \div 50\,\mu$m) with high average power 0.1-1 MW for photochemical research.

Several difficult problems arose during the development of the powerful FEL and they have been conceptually solved in the project.

The electron energy 50 MeV at peak current of more than several amperes is needed for the infrared FEL operation. The average power in the electron beam can reach 100 MW. A race-track microtron was chosen as the FEL driver. The main distinguishing feature of the race-track microtron is the use of energy recuperation.

The main ides of the energy recovery system is the following: the recirculated electron beam is injected into the linac in such a phase that it loses energy to the cavity field. In this mode of operation the energy delivered by the cavity fields to the electron beam on the first pass through the linac is restored to the cavity fields by the electrons on the second pass. Energy recovery of this type was first suggested at Los Alamos by Brau *et al.* [4.20].

An advantage of the energy recovery system which might be just as important as the reduction in size of the rf components is its self-limiting nature. For example, with 50 MeV and 10 mA electron beams the beam power through the wiggler is 500 kW. Not only is this enough beam power to cause a lot of damage to beam line components if it should be steered incorrectly, it is enough beam power that protection from the radiation produced will require substantial amounts of shielding. With a properly designed energy recovered system the small amount of total rf power available guarantees that 500 kW of beam power cannot be dissipated anywhere. The maximum beam power that can be dissipated is clearly limited to the rf power available, and in fact, will certainly be less due to an automatic mismatch between the rf power source and the cavity when the transmission of the recirculation system degrades.

Main parameters of this machine are given in Table 4.1.

The layout of the microtron is shown in Fig. 4.18. The microtron comprises an injector 1, two magnetic systems of a 180°

Table 4.1. The Microtron-Recuperator Parameters

RTM RF wavelength	166.3 cm
Number of RF cavities	20
Number of tracks	4
Energy gain per one RF cavity	0.7 MeV
Injection energy	2.1 MeV
Final electron energy	51 MeV
Final electron energy dispersion	0.45%
Final electron micropulse length	20–100 ps
Final electron peak current	20–100 A
Micropulse repetition frequency	2–45 MHz
Electron average current	4–100 mA

separating bend 2, a common straight section with rf cavities 3 (the section is common to electrons of different energy), magnets for the injection 4 and extraction 5 systems, solenoidal magnetic lenses 6, four separated straight sections with magnetic quadrupole lenses 7, FEL magnetic system 8 placed in the fourth straight section, and a beam dump 9.

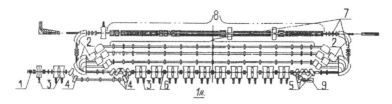

Figure 4.18. The layout of the microtron.

The 300 kV electron gun of the injector generates 1 ns electron bunches at 45 MHz. Having passed through an rf cavity, the bunch is longitudinally compressed in a straight bunching section down to 150 ps and then accelerated up to 2.1 MeV in two rf cavities of the injector. The electrons are injected into the common straight section of the microtron using a 180° magnetic mirror and two identical 65° bending rectangular magnets of opposite sign.

The working voltage amplitude of the RF cavities in the common straight section of the microtron is 700 kV for one cavity and, hence 14 Mev for 20 cavities, with the flight factor taken into account.

The fourth straight section is intended for the FEL magnetic system. At the exit from the FEL magnetic system there are two RF cavities to compensate for the average losses in electron energy in the FEL. Entering the common straight section from the fourth track, but now in a decelerating phase of the rf voltage, the electrons release their energy to the rf system during the passage in the same direction through the same three microtron tracks. After that, the electrons are extracted using the magnets of the extraction system and are directed to the beam dump.

Calculation of the longitudinal and transverse beam dynamics show that the microtron-recuperator can operate in a steady mode with an average current higher than 0.1 A, peak current ~ 100 A with a low transverse emittance of the beam.

One of the key problems for high power laser oscillators are the extremely high light fields inside their optical cavities. Usually lasers operate on modes close to the fundamental one and therefore the light spot size on the cavity mirrors is relatively small. This can lead to mirror damage by intracavity light. In the FEL under consideration this problem is aggravated by the high average light power. The problem was solved by the FEL group in INP in very elegant way. The following concept of radiation extraction was suggested, the so-called 'electron radiation extraction' and has been successfully modeled.

The magnetic system of the FEL consists of four undulators, two dispersive sections and one achromatic bend. The first three undulators and two dispersive sections compose the optical klystron used as a master oscillator. Its optical cavity is formed by two mirrors and it is 79 m long. The undulator period is ≃9 cm, the number of periods in each undulator is 40. The electromagnetic undulator permits to vary the undulator parameter K from 1 to 2.

The master oscillator with the optical cavity is used only for electron bunching and therefore the intracavity power is relatively low. The bunched electron beam goes to the fourth output undulator where it produces powerful coherent radiation directed forward. To avoid damage of the forward mirror under the output radiation the bunched beam enters the output undulator after a small (4 milliradians) angle bend.

The FEL radiation will consist of 10–30 ps pulses, repetition rate 2–45 MHz at a 4–13 μm wavelength. Varying the electron energy from one bunch to another with the round-trip period of the optical cavity, it is possible to modulate the wavelength. Estimates of the coherent radiation power are of the order of MW peak power at 100 A peak current and of few tens of kW average power with a 0.1 A average current.

To increase the peak current of the microtron and to improve electron beam quality a CW photoinjector for the high power FEL is under development [4.21]. It is expected that the photogun beam will be have 10^{-7} cm×rad emittance and less than 1 keV energy spread at 100 mA average current, 300 kV electron energy and 200 ps pulse length. It is planned that the high power FEL facility will be available for users in near future.

Part II. SECONDARY BEAMS

Chapter 5

Interaction of Relativistic Electrons with Matter

5.1 Ionization and Radiative Losses of Electrons

We first discuss in this chapter fast electrons travelling through matter with energies that exceed some orders of magnitude by the mean binding electron energy in an atom, called the *mean ionization potential* $\bar{I}$. For the value of $\bar{I}$ the following empirical relationship is valid

$$\bar{I} = 13.5\, Z \text{ eV}. \tag{5.1}$$

We consider passage through matter of electrons with energies of 0.01-0.1 MeV and more, and it means that we can neglect binding of electrons with nuclei and consider them as free particles in their impact with incoming fast electrons. Electromagnetic interaction between incoming relativistic electrons and the Coulomb field of nuclei leads to radiation of electromagnetic waves, and, therefore, electrons lose in medium their primary energy by ionization and radiative processes.

The cited processes of energy losses are related to the elementary act of interaction of incoming particles in a homogeneous medium. The motion of electrons in an inhomogeneous medium leads to the origination of a distinct form of radiation. First of all it is transition radiation – the production of photons when energetic charged particles cross the interface between two dielectric media. On the other hand, an interaction of electrons with a crystal lattice yields X-rays in two ways: 1) parametric radiation, when the virtual photon field associated with a relativistic electron travelling in a crystal is diffracted by the crystal lattice in the same way that real X-rays are diffracted by crystals, and 2) channeling radiation when radiation results from spontaneous transitions between eigenstates introduced by crystalline fields in the direction transverse to the particle's relativistic motion.

As it follows from the above arguments the main mechanism of the interaction of high energy electrons with matter is the following. A fast electron, passing through matter, 'pushes aside' atomic electrons by its Coulomb field and in such a way it loses its energy. These losses are called *ionization losses* as in each collision the atomic electron is raised to the continuum of states and the atom is ionized.

Although the mechanism of ionization losses for electrons is the same as for other charged particles, two peculiarities should be borne in mind:

(i) For a heavy charged particle it is usually supposed that in a collision with an electron an incoming particle is slightly deflected from its path and its trajectory is a practically rectilinear one. It is not so for electrons, which can be considerably deflected from their primary direction of motion, and, besides, electromagnetic radiation can be originated in its impacts. The first effect means its motion is not rectilinear, the second means that *radiative losses*, i.e., losses by electromagnetic radiation, can be significant.

(ii) Collisions between identical particles involve an energy exchange phenomenon which has, therefore, to be taken into account for electron-electron collisions.

Taking into consideration these effects Bethe obtained the following formula for electron (positron) energy losses:

$$-\frac{dE_e}{dx} = \frac{2\pi e^4 n}{mv^2}\left[\ln\frac{mv^2 E_k}{2\bar{I}^2(1-\beta^2)} - \right.$$

$$\left. (2\sqrt{1-\beta^2} - 1 + \beta^2)\ln 2 + 1 - \beta^2 + \frac{1}{8}(1 - \sqrt{1-\beta^2})^2\right], \quad (5.2)$$

where $E_k = mc^2(\gamma - 1)$ is the relativistic kinetic energy.

In the ultrarelativistic limit $E \gg mc^2$ formula (5.2) has the form

$$-\frac{dE}{dx} = \frac{2\pi e^4 n}{mc^2}\left[\ln\frac{E^2}{2\bar{I}^2\sqrt{1-\beta^2}} + \frac{1}{8}\right]. \quad (5.3)$$

We will now consider radiation losses. A charged particle passing through the Coulomb field of a nucleus with the charge Ze

is deflected, i.e. it acquires an acceleration w, that is accompanied by radiation. This radiation is called *bremsstrahlung*. If the charge of the particle equaals e, and where m is the mass, then

$$|w| = \frac{F}{m} = \frac{e\mathcal{E}}{m} = \frac{Ze^2}{mr^2} \, , \qquad (5.4)$$

where $\mathcal{E}$ is the intensity of the electrostatic field, and r is the distance from the center of nucleus called *impact parameter*.

The intensity W of bremsstrahlung, i.e. the quantity of energy emitted per secjnd for the particle moving with acceleration w in a nonrelativistic limit is determined by

$$W = \frac{2}{3}\frac{e^2}{c^3}|w|^2 = \frac{2}{3}\frac{Z^2e^6}{m^2c^3r^4} \, . \qquad (5.5)$$

Here c is the velocity of light. Thus, radiative loss varies inversely with the mass of the particle squared, and proportional to the charge of the scattering center squared. Consequently, radiative losses can be significant only for electrons, not for heavy particles. For instance, radiative losses for electrons are $(m_p/m_e)^2 \simeq 3\cdot 10^6$ times as much as for protons. Then, as the major part of ionization losses are due to collisions of incoming particles with atomic electrons, the radiative losses are *vice versa* due to collisions with nuclei. In fact, the radiation intensity from the collision of a particle with a nucleus is Z^2 times as high as from collisions with electrons, but the number of electrons is only Z times as high as nuclei. It is why substances with great Z are used as bremsstrahlung targets. These qualitative estimations are not changed by taking into consideration quantum and relativistic features.

On the basis of quantum electrodynamics Bethe and Heitler have obtained the following formula for radiative electron losses

$$-\frac{dE_e}{dx}\bigg|_{\text{rad}} = nE_eZ^2\Phi(E_e). \qquad (5.6)$$

Here n is the atomic density of the media, $\Phi(E)$ is a linear function of $\ln E$. So, as it was shown above qualitatively, radiative losses are proportional to the charge of the target nucleus squared and the energy of the incoming electron. Hence, an efficiency of electron kinetic energy to radiation increases with electron energy and the charge of target nuclei.

It is also necessary to lay emphasis on the following peculiarities of bremsstrahlung:

(i)   The energy spectrum is continuous up to the upper end that equals the kinetic energy of the bombarding electron.

(ii) The radiation is concentrated in a narrow forward angle of the order of $1/\gamma = mc^2/E$.

(iii) The radiation power (the energy flux through $1\,cm^2$ per sec) is approximately proportional to electron energy cubed.

The ionization and radiation losses are connected by the relation

$$\frac{(dE_e/dx)_{\text{rad}}}{(dE_e/dx)_{\text{ion}}} \simeq ZE. \qquad (5.7)$$

If the energy is measured by MeV then

$$\frac{(dE_e/dx)_{\text{rad}}}{(dE_e/dx)_{\text{ion}}} \simeq \frac{EZ}{800}. \qquad (5.8)$$

One can conclude from this that the radiation losses in normal water ($\overline{Z} = 8$) become equal to the ionization losses at energy $E \simeq 100\,MeV$, but this happens at only $E = 10\,MeV$ for lead. The energy corresponding to equality of ionization and radiation losses is called the *critical energy*, and the distance x_0, at which the energy of the electron is $1/e$ as large due to the radiation losses, is called the *radiation length*. It equals $36\,g/cm^2$ for air and water, $24\,g/cm^2$ for Al, and about $6\,g/cm^2$ for Pb.

In practice bremsstrahlung is produced by stopping fast electrons in thick targets when multiple emission of energy by the same electron and selfabsorption of the γ-quanta produced in the target is essential. We shall consider all practical aspects of bremsstrahlung later.

5.2 Multiple Scattering of Electrons

Passing of a narrow electron beam through matter leads to distortion of its angle, as although left and right deflection occurs with equal probability the mean square angle is not zero.

As angular dispersion of an electron beam is a result of a lot of independent accidental processes the angular distribution can be considered to have Gaussian form

$$\Phi(\vartheta) = A \exp(-\vartheta^2/\overline{\vartheta^2}). \qquad (5.9)$$

The mean square angle of deflection $\overline{\vartheta^2}$ is proportional to the number of collisions N equals the ratio of target thickness L to the free path length l.

If $\overline{\vartheta_1^2}$ is mean square angle for single collision then

$$\overline{\vartheta^2} = \overline{\vartheta_1^2}N = \overline{\vartheta_1^2}\frac{L}{l} = n\sigma L\overline{\vartheta_1^2} = \overline{\theta^2}L, \qquad (5.10)$$

where σ is the scattering cross section, n is the number of nuclei in unit volume, and $\overline{\theta^2}$ is the mean square deflection angle over the unit path, i.e. $\overline{\theta^2} = n\sigma\overline{\vartheta_1^2}$. According to this definition $\overline{\vartheta_1^2} = \frac{1}{\sigma}\int \sigma(\vartheta)\vartheta^2\,d\Omega$, and so $\overline{\theta^2} = n\int \sigma(\vartheta)\vartheta^2\,d\Omega$. Thus, for Rutherford scattering we get

$$\overline{\vartheta^2} \simeq 2\pi n\frac{Z^2}{E^2}\ln\frac{\vartheta_{\max}}{\vartheta_{\min}}. \qquad (5.11)$$

This formula is valid in relativistic, as well as in nonrelativistic cases.

Consider two limit cases: $Ze^2/\hbar v \gg 1$ and $Ze^2/\hbar v \ll 1$. In the first case scattering can be considered classical for all angles. Really, it means that the angle of Rutherford scattering ϑ_{cl} that is $2Ze^2/p\rho v$ has to be much more than the diffraction angle ϑ_{d}. Let us estimate the angle of diffraction. If we fix the impact parameter ρ then in accordance with the uncertainty principle

$$\Delta p \sim \hbar/\rho \quad \text{i.e.} \quad \vartheta_{\mathrm{d}} \sim \frac{\Delta p}{p} \sim \frac{\hbar}{p\rho}. \qquad (5.12)$$

So, the criterion of applicability of classical mechanics for a scattering problem is

$$\vartheta_{\mathrm{cl}} = \frac{2Ze^2}{p\rho v} \gg \frac{\hbar}{p\rho} \quad , \text{i.e.} Ze^2/\hbar v \gg 1. \qquad (5.13)$$

Now let us estimate the classical scattering angle:

$$\vartheta_{\mathrm{cl}} \sim \frac{F_\perp \Delta t}{p} \sim \frac{\partial U}{\partial \rho}\frac{\Delta t}{p} \sim \frac{U(\rho)}{p}\frac{\Delta t}{\rho} \sim \frac{U(\rho)}{pv} \sim \frac{U(\rho)}{E}. \qquad (5.14)$$

Here we use an estimation $\partial U/\partial \rho \sim U/\rho$, considering ρ as the characteristic length on which the function $U(\rho)$ is essentially changed, and E is the particle energy.

The maximum scattering angle will be in the case when the impact parameter equals the nuclear radius R, i.e. $\vartheta_{\max} \sim U(\rho)/E = \frac{Z/R}{E}$. The forward scattering in classical case will occur only if the particle energy E is more the interaction potential $U = Z/R$, i.e. if $Z/RE > 1$ and $\vartheta_{\max} \sim 1$. The minimum deflection angle is determined by the atomic size a: $\vartheta_{\min} = Z/aE$.

For heavy atoms $a \sim Z^{-1/3}$ (the Thomas-Fermi model), and thus $\vartheta_{min} \sim Z^{4/3}/E$.

Hence for electrons with energy $E < Z/R$ we have

$$\overline{\vartheta^2} \sim 2\pi n \frac{Z^2}{E^2} \ln \frac{E}{Z^{4/3}}. \tag{5.15}$$

Under the condition $Ze^2/\hbar v \ll 1$ the angles ϑ_{min} and ϑ_{max} have to be determined in the following way:

(i) ϑ_{min} is the angle of diffraction on atomic radius, i.e $\vartheta_{min} \sim \lambda/a$, and as $a \sim Z^{-1/3}$ and de Broglie wavelength $\lambda = \hbar/p$ we get $\vartheta_{min} \sim Z^{1/3}\hbar/p$;

(ii) ϑ_{max} is the angle of diffraction on a nucleus, i.e. $\vartheta_{max} \sim \lambda/R \sim \hbar/pR$ at $\lambda/p < 1$ and $\vartheta_{max} \sim 1$ at $\lambda/R > 1$.

So, in the case $Ze^2/\hbar v \ll 1$ we get

$$\overline{\vartheta^2} = 2\pi n \frac{Z^2}{E^2} \times \begin{cases} \ln a/R & \text{if } \lambda/R < 1, \\ \ln a/\lambda & \text{if } \lambda/R > 1. \end{cases} \tag{5.16}$$

5.3 Transition Radiation

Only relatively late — after discovery of the Vavilov-Cherenkov effect and the development of its theory by Frank and Tamm [5.1] — did it become clear that a charge moving uniformly along a straight line can emit waves. The Vavilov-Cherenkov (in what follows V.-Ch.) radiation, is well known to takes place if a charge (or another source which does not have proper oscillations such as, for example, an electric dipole or magnetic dipole) moves at a constant velocity v, larger than the phase velocity of light in a transparent medium $c_{ph} = c/n(\omega)$; $n(\omega)$ is here the refractive index for the frequency ω. The angle θ of emission (i.e. the angle between v and the wave vector k) is determined by the known relation

$$\cos \theta = c/n(\omega)v. \tag{5.17}$$

This condition is simply a kinematic one since the fulfillment of (5.17) means that waves of any nature with phase velocity c_{ph}, emitted along the source trajectory are in phase with each other on the corresponding conical surface (see Fig. 5.1). It is natural therefore that relation (5.17) written in the form $\cos \theta = c_{ph}/v$ was known long ago in acoustics as a condition which determines the angle of the so-called Mach cone, i.e. a conical wave, emitted by a supersonic source (in acoustics c_{ph} is obviously equal to the sound velocity).

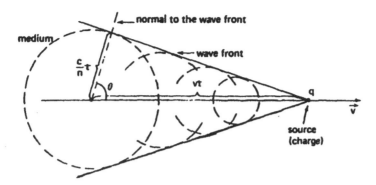

Figure 5.1. Formation of Vavilov-Cherenkov radiation (ct/n is the length of the light path during the time t, vt is the length of the source path during the same time).

We will be interested in describing the emission of a source (a charge, multipole etc.) moving uniformly and rectilinearly and the V.-Ch. emission takes place in this case. Moreover, in the case of a medium homogeneous in space and constant in time the V.-Ch. emission is the only kind possible. But in an inhomogeneous medium, in a medium with properties changing in time or in a vacuum close to such a medium the situation is quiet different. Namely, under these conditions transition radiation is possible. In the full sense of the word transition radiation is emission appearing for a charge (or other source without proper oscillations) moving with constant velocity under inhomogeneous conditions, i.e. in an inhomogeneous medium, in a medium changing over time or close to such medium.

So, consider a charge, moving with a constant velocity

$$v < c/n, \tag{5.18}$$

when V.-Ch. emission does not occur. In the case of a vacuum ($n = 1$) there will be no emission for any $v = $ constant (in the absence of boundaries). To have emission in a vacuum the charge (or multipole) should be accelerated, or in other words the parameter v/c should be changing. In the presence of a medium this parameter has the form $v/c_{ph} = vn/c$, i.e. it is equal to the ratio of the particle velocity to the light phase velocity $c_{ph} = c/n$. The most important point is that in the presence of a medium the parameter vn/c can change both due the variation of the velocity v and due to the change along the charge trajectory of the phase velocity $c_{ph} = c/n$ which varies according to the corresponding change in the refractive index n. The emission due to the change of n in the parameter vn/c for $v =$ constant is just, generally speaking the transition emission. The simplest problem of this type is when a charge crosses the boundary of two media (particularly the boundary of a vacuum and a medium). This simplest type of transition radiation was first considered by Frank and Ginzburg in 1944 [5.2].

The explanation of transition radiation as a result of a change in the parameter vn/c given above is in some sense a formal one and in reality needs an insight into the theory of emission in a medium. Therefore the most simple explanation of the transition radiation in the case when the charge crosses a boundary is worth remembering. It is well known that the electromagnetic field of the charge in one (the first) of the two media can be constructed as a superposition of the field of the charge 'itself' and the field of an 'image' charge, which moves in the other (second) medium. At the moment when the charge is on the boundary from the 'point of view' of the field in the first medium the charge and its 'image' partially 'annihilate' each other, which creates radiation. The simplest is the case of normal (to the surface of the boundary) transit of a charge from vacuum into an ideal mirror. In this case one can say that at the boundary the charge and its image annihilate each other completely or one can say better that the charge and its image stop at the boundary. The radiation appearing in vacuum should be the same as that appearing from the charge q and its image $-q$ stopped instantaneously at the boundary (see Fig. 5.2).

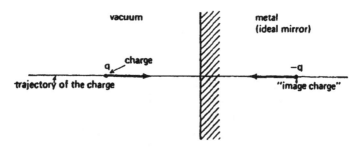

Figure 5.2. Formation of transition radiation by a charge crossing a boundary between a vacuum and metal.

From this picture it is obvious that transition radiation should appear in the case when the source crosses the boundary between any media with different electric properties (different dielectric constants, for example). For high-energy particles, and particularly relativistic electrons, the case when the particle passes through the medium and enters the vacuum is of practical interest. From the point of view of the theory, this case is equivalent to the previous one since the corresponding result can be simply found by the change of v to $-v$. On the other hand, there is clearly no symmetry in the intensity of radiation and the intensities of forward and backward radiation (in the case when the particle enters the medium and in the case when the particle leaves the medium to the vacuum) are different and in some cases differ from each other very significantly. Particularly in the intensity of the forward radiation (in the case of a particle transit from a medium to the vacuum) the higher harmonics of radiation are present. In the usual condensed (solid) media the forward radiation of relativistic particles can be spread in frequencies up to the X-ray domain. Accordingly the integrated (over angle and frequencies) intensity in the forward direction is enlarged and is in the simplest case proportional to $1/\sqrt{1 - v^2/c^2} = \mathcal{E}/Mc^2$, where $\mathcal{E}$ is the total energy of an emitting charge of mass M. This important point was cleared up only in 1959 by Garibyan [5.3] and by Barsukov [5.4].

The photon production for a single interface is small; however, by stacking a number of foils, the yield can be greatly increased. In most applications, individual foils separated by a vacuum are used to reduce re-absorption of the X-rays in the medium.

The photon production from transition radiation is closely related to the thickness of an individual foil, not only because there is re-absorption of the emitted radiation in the foils themselves, but also because a minimum thickness (known as the formation length) is needed for photon production. Re-absorption can be minimized by making the foils as thin as possible; however, if they are made thinner than the formation length, the photon production will drop. Thus, there is an optimum foil thickness that balances production with re-absorption, giving a maximum photon yield. Foe soft X-rays, the foil thicknesses usually used are between 0.5 and 5 μm.

In general, the radiator will be of thin foils of thickness l_2 and plasma frequency ω_2 ($\omega^2 = 4\pi n e^2/m$, n is the electron density of the medium) separated by either a gas or vacuum of thickness l_1 and plasma frequency ω_1. For the usual case, when $l_1 \gg l_2$ and $\omega_2 \gg \omega_1$, the radiation is emitted at frequencies $< \gamma\omega_2$, where $\gamma = E/mc^2 - E/0.511$, and E is the electron-beam energy in MeV. This frequency represents a cutoff above which the radiation falls dramatically. Since the plasma frequency of a material is proportional to the square root of its density, the cutoff frequency is proportional to the square root of the foil density. For beryllium foils, $\omega_2 = 24.5\,eV$, and a γ of 50 to 100 is needed for adequate photon production at 1.5 keV.

The frequency spectrum of transition radiation emitted by a single electron upon perpendicular traversal through a single-foil interface has been calculated by Ginzburg and Frank [5.2] and by Garibyan [5.3] (summary and a detailed discussion of the theoretical results is given in [5.5]). These calculations show that for highly relativistic particles most of the radiation is emitted in the X-ray region. With $\epsilon_{1,2} = 1 - \omega_{1,2}^2/\omega^2$ (where $\omega_{1,2}$ are plasma frequencies of the two media and equal $4\pi n_{1,2} e^2/m$, and $n_{1,2}$ is the electron density of the media), relatively simple expressions have been found for the differential and the total radiation intensities:

$$\frac{d^2 S_0}{d\theta d\omega} = \frac{2\alpha\hbar\theta^3}{\pi} \left[\frac{1}{1/\gamma^2 + \theta^2 + \omega_1^2/\omega^2} - \frac{1}{1/\gamma^2 + \theta^2 + \omega_2^2/\omega^2} \right]^2, \quad (5.19a)$$

$$\frac{dS_0}{d\omega} = \frac{\alpha\hbar}{\pi} \left[\left(\frac{\omega_1^2 + \omega_2^2 + 2\omega^2/\gamma^2}{\omega_1^2 - \omega_2^2} \right) \ln \left(\frac{1/\gamma^2 + \omega_1^2/\omega^2}{1/\gamma^2 + \omega_2^2/\omega^2} \right) - 2 \right], \quad (5.19b)$$

$$S_0 = \int \int \left(\frac{d^2 S_0}{d\theta d\omega} \right) d\theta d\omega = \frac{\alpha\hbar}{3} \frac{(\omega_1 - \omega_2)^2}{\omega_1 + \omega_2} \gamma, \quad (5.19c)$$

where ω = photon frequency, θ = emission angle of the photon with respect to the particle trajectory, $\alpha = e^2/\hbar c = 1/137$ is the fine structure constant.

Transition X-rays are emitted into a narrow conical beam directed forward (see Fig. 5.3). The cone angle θ is approximately $1/\gamma$, with a width that also is about $1/\gamma$ as it follows from (5.19a).

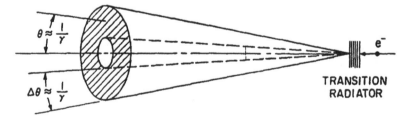

Figure 5.3. A schematic diagram of the cone of emission produced by transition radiation showing the peak emission angle and range of emission angles.

Equation (5.19b) describing the differential energy spectrum can be broken up into essentially three regions having respectively, constant, logarithmic, and power-law dependence on ω and γ:

$$\frac{dS_0}{d\omega} \approx \begin{cases} \dfrac{2\alpha\hbar}{\pi}\left(\ln\dfrac{\omega_1}{\omega_2} - 1\right), & \omega < \gamma\omega_2 & (5.20a) \\[3mm] \dfrac{2\alpha\hbar}{\pi}\ln\dfrac{\gamma\omega_1}{\omega}, & \gamma\omega_2 < \omega < \gamma\omega_1 & (5.20b) \\[3mm] \dfrac{\alpha\hbar}{6\pi}\left(\dfrac{\gamma\omega_1}{\omega}\right)^4, & \gamma\omega_1 < \omega. & (5.20c) \end{cases}$$

At frequencies exceeding a 'cutoff' frequency $\omega = \gamma\omega_1$, the radiation intensity drops rapidly to very small values. An example of the spectrum (5.19b) is shown in Fig. 5.4.

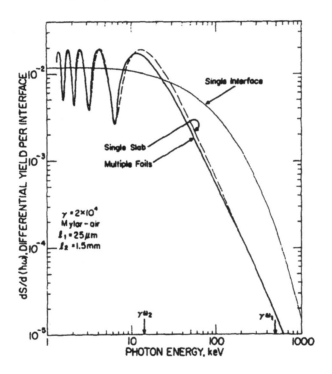

Figure 5.4. Comparison of $dS/d(\hbar\omega)$, the computed differential yield (keV/keV) per interface, as a function of $\hbar\omega$, the photon energy, for different Mylar-air radiators. For a single interface (thin line), $dS/d(\hbar\omega) = dS_0/d(\hbar\omega)$; for a single slab (thickness $l_1 = 25\ \mu$m, dashed line), $dS/d(\hbar\omega) = \frac{1}{2}dS_1/d(\hbar\omega)$; and for multiple foils (thickness $l_1 = 25\ \mu$m and spacing $l_2 = 1.5$ mm, heavy line), $dS/d(\hbar\omega) = (1/2N)dS_N/d(\hbar\omega)$. The particle energy is $\gamma = 2 \cdot 10^4$ (from [5.5]).

The most apparent feature of transition radiation seems to be its dependence on the Lorentz factor γ of the particle: the total intensity (Eq. (5.19c)) increases linearly with γ, essentially due to the fact that the photon spectrum becomes harder with increasing γ. It should be noted, however, that measurements with detectors which are only sensitive in a limited frequency interval would, in general, not follow a linear γ dependence, but develop with logarithmic behaviour (Eq.(5.20b)) and eventually reach saturation when $\gamma > \omega/\omega_2$ – Eq.(5.20a). The logarithmic dependence would persist for large values of γ only if one of the two media was a vacuum $(\omega_2 = 0)$.

Complications arise due to the fact that the transition radiation generated at a single interface can in general not be observed. In practice, the particle traverses at least two interfaces of a slab of material. Furthermore, since the total radiation yield per interface is very small $(S_0 \approx 10^{-2}\gamma\,\mathrm{eV}$ – Eq.(5.19c)), the effect must be enhanced for experimental studies. A radiator may be chosen in which the particle traverses many interfaces, for instance, a stack of thin foils of low-Z material, stretched in air at constant spacings. In this case, interference effects between the individual interfaces of the radiator must be taken into account, leading to a modulation of the frequency distribution obtained from a single interface, and, for high values of γ, to a saturation of the yield. The differential spectrum of X-ray transition radiation for a periodic radiator of N foils with thickness l_1 and spacing l_2 is given by

$$\frac{d^2 S_N}{d\theta d\omega} = \frac{d^2 S_0}{d\theta d\omega} 4 \sin^2\left(\frac{l_1}{Z_1}\right) \frac{\sin^2[N(l_1/Z_1 + l_2/Z_2)]}{\sin^2(l_1/Z_1 + l_2/Z_2)}, \qquad (5.21a)$$

where Z_1 and Z_2 are the 'formation zones' for the two media:

$$Z_{1,2} = \frac{4c}{\omega}\left(\frac{1}{\gamma^2} + \theta^2 + \frac{\omega_{1,2}^2}{\omega^2}\right)^{-1}. \qquad (5.21b)$$

Physically, the formation zone is the distance along the particle trajectory in a given medium after which the separation between particle and generated photon is of the order of the photon wavelength.

Expression (5.21a) can be obtained simply by linear superposition of the single-interface expressions (5.19).

Let us discuss the consequences of Eq.(5.21a) in some detail. For the case $N = 1$, we obtain the radiation yield of a single slab of material:

$$\frac{d^2 S_1}{d\theta d\omega} = \frac{d^2 S_0}{d\theta d\omega} 4 \sin^2\left(\frac{l_1}{Z_1}\right). \qquad (5.22)$$

From an experimental view point, it is interesting to determine the condition for which the yield from a single slab is comparable to that from two single interfaces. To this end, we consider a detector which has frequency resolution $\Delta\theta$, and uniform response over $\Delta\omega$ and $\Delta\theta$. The *observed* differential spectrum is then the spectrum of Eq. (5.22), averaged over $\Delta\theta$ and $\Delta\omega$. If the thickness of the foil exceeds the formation zone for the foil material,

$$l_1 \gg Z_1(\theta,\omega), \tag{5.23}$$

then the term $\sin^2(l_1/Z_1)$ in Eq.(5.22) oscillates rapidly as compared to the range in which $d^2S_0/d\theta d\omega$ varies considerably. If it also oscillates rapidly as compared to the resolution of the detector, then the average over $\Delta\theta$ and $\Delta\omega$ is

$$\left\langle \frac{d^2S_1}{d\theta d\omega} \right\rangle_{\Delta\theta,\Delta\omega} = \frac{1}{\Delta\theta\Delta\omega} \int_{\Delta\theta}\int_{\Delta\omega} \left(\frac{d^2S_1}{d\theta' d\omega'} \right) d\theta' d\omega' =$$

$$\left\langle \frac{d^2S_0}{d\theta d\omega} \right\rangle_{\Delta\theta,\Delta\omega} \left\langle 4\sin^2 \frac{l_1}{Z_1} \right\rangle_{\Delta\theta,\Delta\omega}, \tag{5.24}$$

and since

$$\langle \sin^2(l_1/Z_1) \rangle_{\Delta\theta,\Delta\omega} \approx \frac{1}{2} \tag{5.25}$$

the observed yield is twice that of a single interface.

Returning to the general form for a periodic radiator (Eq.(5.21)), we consider now the case $l_2 > Z_2(\theta,\omega)$, i.e., the spacing of the foils exceeds the formation zone of the gap material. In this case, the period of the term

$$\frac{\sin^2[N(l_1/Z_1 + l_2/Z_2)]}{\sin^2(l_1/Z_1 + l_1/Z_2)}$$

is very small compared to the range in which $d^2S_0/d\theta d\omega$ varies. If it is also small compared to the detector resolution, then we may replace the last term in Eq.(5.21a) by its average:

$$\left\langle \frac{\sin^2[N(l_1/Z_1 + l_2/Z_2)]}{\sin^2(l_1/Z_1 + l_2/Z_2)} \right\rangle_{\Delta\theta,\Delta\omega} \approx N. \tag{5.26}$$

Thus, the observed yield in this case is well approximated by N times the radiation yield of a single slab.

In the opposite case, i.e., when the foil spacing l_2 is smaller than the formation zone Z_2, i.e., $l_2 < Z_2(\theta,\omega)$, the approximation (5.26) cannot be used even if $l_1 \gg Z_1$. In general, for $l_2 < Z_2$, a reduction of the total yield per interface occurs as compared to N single slabs, until, as the foil spacing approaches zero $(l_2 \to 0)$, the radiation yield corresponds to that of a single slab with thickness Nl_1.

To obtain the total intensity of radiation emitted from N slabs it is necessary to integrate Eq.(5.21a) over angle and photon energy, and the result is plotted, for comparison with single interface prediction, in Fig. 5.4. Inspection of the figure leads to the following qualitative conclusions:

a) The energy spectrum of transition radiation for a periodic radiator is characterized by oscillations of sizeable amplitude around the spectrum for a single interface. Maxima of intensity occur (below $\omega = \gamma\omega_1$) at characteristic frequencies $\omega_{m,r}$:

$$\omega_{m,r} = \frac{(\omega_1^2 - \omega_2^2)l_1 l_2}{2\pi c(l_1 + l_2)[(2rl_2)/(l_1 + l_2) - 2m + 1]}, \qquad (5.27)$$

where m, r are integer numbers.

b) For the case shown in Fig. 5.4, the complication introduced by the expression for a periodic medium over the single slab formula does not result in a major difference in the calculated yield. This is due to the fact that in this case the condition $l_2 \geq Z_2$ is fulfilled, i.e., the air formation zone does not significantly exceed the foil spacing.

c) Most importantly, with increasing particle energy the radiation spectrum of a periodic radiator does not become harder at the same rate as the single-interface radiation. As a result, the linear γ-dependence of the total radiation yield predicted by Eq. (5.19c) will not persist for periodical radiators; rather, as shown in Fig. 5.5, saturation will set in at large γ values. This saturation exists even in the case of a single slab. The 'saturation energy' γ_s is defined as

$$\gamma_s = \frac{1}{4\pi c}\left[(l_1 + l_2)\omega_1 + \frac{1}{\omega_1}(l_1\omega_1^2 + l_2\omega_2^2)\right]. \qquad (5.28)$$

The frequency spectrum of transition radiation as discussed thus far characterizes the X-ray yield generated in the radiator without considering reabsorption. However, reabsorption does strongly suppress the radiation intensity emerging from a radiator at low X-ray energies. This is demonstrated in Fig. 5.6 where the net spectrum computed for a typical radiator is shown.

Figures 5.4 and 5.6 demonstrate that most of the generated radiation occurs with frequencies around the last maximum in the differential frequency spectrum. This effect is enhanced by

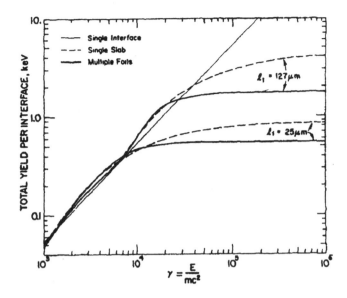

Figure 5.5. Comparison of the total generated yield per interface, S, as a function of the particle energy $\gamma = E/mc^2$ for single interface (thin line), single slab (dashed line), and multifoil radiator (heavy line). The yield is computed for Mylar-air interfaces, with foil thickness $l_1 = 25\,\mu m$ and $127\,\mu m$, and spacing $l_2 = 1.5\,mm$ (from [5.5]).

the suppression of low X-ray frequencies due to absorption (see Fig. 5.6). As saturation (with increasing γ) is approached, this maximum frequency will be

$$\omega_{max} \approx \frac{l_1\omega_1^2}{2\pi c} \quad \text{if} \quad \omega_1 \gg \omega_2 \quad \text{and} \quad l_2 \gg l_1, \quad (5.29)$$

i.e., ω_{max} is roughly proportional to the mass density of the individual radiator foils, and it does not depend on the particle energy. It is interesting to note that radiator thickness l_1, frequency ω_{max} (Eq. (5.29)), and formation zone Z_1 are closely related: the formation zone Z_1, evaluated for $\omega = \omega_{max}$ (and $\theta \approx 1/G$), approximately equals the thickness l_1, i.e., $Z_1(\omega_{max}) \approx l_1$.

When the intensity varies rapidly with both photon energy and angle, the radiation maximum is difficult to resolve. These variations are averaged when detector has low resolution in both solid angle and energy. In addition, the angular distribution is broadened because of both the finite electron-beam dimensions and multiple scattering. In such a case the intensity detected will be

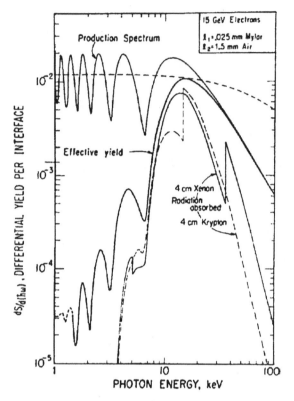

Figure 5.6. Comparison of the generated differential yield of a multifoil radiator (production spectrum) with the yield emerging after taking absorption in the radiator into account (effective yield). The strong suppression of low X-ray energies due to absorption should be noted. The lower curves indicate the fraction of the effective yield that will be absorbed in detectors of 4-cm xenon (solid line) or 4-cm krypton (dashed line) (from [5.5]).

proportional to the factor M

$$M = \frac{1 - \exp(-N\sigma)}{1 - \exp(-\sigma)}. \qquad (5.30)$$

Here $\sigma = \mu_1 l_1 + \mu_2 l_2$ where $\mu_{1,2}$ are the linear absorption coefficients of the spacing and foil media, respectively. We see from Eq. (5.30), that when $N\sigma \gg 2$ the asymptotic value for M is $2/\sigma(\omega)$, and that beyond $2/\sigma$ foils the radiation intensity cannot be increased significantly by adding more foils. In fact, the number of foils N is approximately 10 to 100.

5.4 Parametric Radiation

We refer to the parametric X-rays (PX) which originate when a relativistic charged particle interacts with a crystal. As it was shown in Sec.5.3, this radiation is due to the periodic character of the crystal's dielectric constant. The optical parametric radiation was first described by Fainberg and Thiznyak [5.6]. They considered the radiation from an uniformly moving electron in a medium with a simply periodic dielectric constant. Analogous radiation in a thin triply periodic crystal was considered by Ter-Mikaelyan [5.7[. But true PX occurs in a crystal with thickness L, greater than the X-ray extinction length, which is

$$KL|n-1| \geq 1 \quad , \tag{5.31}$$

where K is the emitted photon wave number and n is the crystal refraction index. In this case the emitted photon state in the crystal is modified in an essential way. The importance of such a modification was first mentioned by Baryshevsky [5.8].

Consider qualitatively the origin of the phenomenon following to Baryshevsky and Feranchuk [5.9]. Let us suppose that an ultrarelativistic electron $(E \gg mc^2)$, enters a crystal. Here E is the energy and m is the mass of the particle. The crystal is arbitrarily oriented with respect to the particle velocity v. In order to give a simple analysis of the PX formation we recall that the relativistic charged particle's electromagnetic field can be represented as a superposition of pseudophotons whose properties are close to those of real photons [5.10]. The pseudophotons have angular spread $\Delta\theta \leq mc^2/E$ and their spectral distribution is determined as follows

$$n(\omega) = \frac{2}{\pi\omega} \ln\left(\eta\frac{E}{\omega}\right), \quad \omega \ll E, \tag{5.32}$$

where ω is the pseudophoton frequency and η has a value of the order of unity.

From this point of view, the electromagnetic interaction of the relativistic electron with the crystal is equivalent to the interaction of a photon beam (angular spread $\Delta\theta$ and spectrum I) with the crystal (see Fig. 5.7). PX can then be considered to be the result of X-pseudophoton diffraction in the crystal. The pseudophoton diffraction as well as the diffraction of external X-rays is a coherent process, i.e., the probability of this process is in direct proportion to the square of the crystal length. Therefore under certain fixed conditions the PX intensity may essentially exceed the intensity of the pseudophoton incoherent scattering which leads to bremsstrahlung.

Because of the pseudophoton diffraction the PX travels both along of the electron velocity and at large angles relative to it. These angles are determined by the crystal reciprocal lattice vectors (for example, the parametric radiation in the $-\vec{v}$ direction is possible). This angular distribution is the most important feature inherent to PX which enables us to separate this radiation from other secondary processes, due to the ultrarelativistic particle interaction with the crystal. The point is that the particles formed in the secondary processes (photons, electron-positron pairs and so on) leave the crystal at small angles ($\sim mc^2/E$) relative to $\vec{v}$.

In order to consider the interaction of the pseudophotons with the crystal one can use the results of the theory of the diffraction of X-rays and resonant γ-radiation. In the case $E \gg mc^2$ the pseudophoton momentum can be written as

$$\vec{K} = \omega \frac{\vec{v}}{v^2} \approx \omega\vec{v}. \tag{5.33}$$

Those pseudophotons whose momenta satisfy the Bragg condition

$$(\vec{K} + \tau)^2 \approx K^2 \tag{5.34}$$

are diffracted by the crystal. Such a process leads to parametric X-radiation in the $(\vec{K} + \tau)$ direction. Essentially, the photons are emitted from the crystal at an angle which does not depend on the energy E. This angle is defined by the particle's velocity orientation relative to the crystallographic plane (Fig. 5.8). Here τ is the reciprocal lattice vector of the crystal.

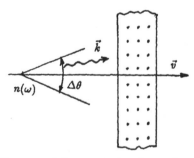

Figure 5.7. Diagram of the spectral and angular spread of pseudophotons.

As a result, the condition of the Vavilov-Cherenkov radiation

$$1 - vn(\vec{K}, \omega)\cos\theta = 0 \tag{5.35}$$

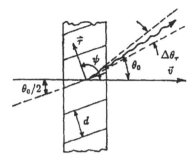

Figure 5.8. Geometry of an experiment to observe the parametric X-rays.

is satisfied in the crystal, in contrast with the homogeneous medium, where $n < 1$ for X-rays. Thus, the V.-Ch. radiation under the diffraction condition (5.34) is the parametric X-rays.

Depending on τ, the emitted photons propagate both at an angle smaller than $\pi/2$ relative to $\vec{v}$ (the Laue case) and at an angle greater than $\pi/2$ (the Bragg case). The pseudophoton spectrum defined by Eq. (5.32) is continuous and therefore the condition is fulfilled for all reciprocal lattice vectors simultaneously. The intensity of PX in the direction of $(\vec{K} + \tau)$ depends on the structure factor $F(\tau)$ determining the probability of X-ray coherent scattering with momentum transfer $\vec{\tau}$. An analogous situation takes place when a polychromatic X-ray beam is diffracted by a crystal.

As we mentioned above, the pseudophotons have an angular spread $\Delta\theta \sim mc^2/E$. As a consequence PX propagates in the direction of $\vec{K} + \tau \approx \omega\vec{v} + \tau$ in a cone with angular divergence $\Delta\theta$. Therefore, lateral spots have to appear on the X-film placed to the right or to the left of the crystal. The distribution of these spots does not depend on the particle energy and coincides with the totality of reflexes formed when an X-ray beam with the spectrum (5.32) is diffracted by the crystal. The frequency of PX photons in each reflection is concentrated near its own frequency ω_B, which is

$$\omega_B = \frac{\tau^2}{2|\tau_z|} = \frac{\tau}{2\sin\theta_B}. \tag{5.36}$$

Figure 5.9 shows the distribution of the most intense reflections of PX for three mutually perpendicular directions of the particle incidence in crystals with lattices of the diamond type, i.e., body centered cubic lattices (such crystals have the maximum density of atoms and therefore possess the largest intensity of PX).

Figure 5.10 shows the typical form of one such reflection.

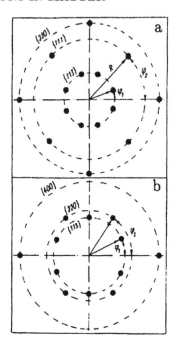

Figure 5.9. The distribution of the most intensity reflexes of the parametric X-rays in the crystals with lattice of diamond type; $R \sim \theta_B$: a) the particle velocity is directed along the axis $\langle 100 \rangle$; $\varphi_1 = 18.43°$; $\varphi_2 = 45.53°$; b) the same for axis $\langle 110 \rangle$; $\varphi_1 = 19.5°$; $\varphi_2 = 64.3°$ (from [5.11]).

The theoretical expressions for PX have been summarized by Feranchuk and Ivashin [5.11]. According to these authors, the angular distribution of PX intensity can be expressed as

$$\frac{\partial^2 N}{\partial \theta_x \partial \theta_y} = \sum_{n=0}^{\infty} \frac{e^2}{4\pi} \omega_B^{(n)} L_a \left[1 - \exp\left(-\frac{L}{L_a}\right)\right] \times$$

$$\frac{|g_\tau(\omega_B^{(n)})|^2}{\sin^2 \theta_B} \frac{\theta_x^2 \cos^2 \theta_B + \theta_y^2}{[\theta_x^2 + \theta_y^2 + \theta_{\rm ph}^2]^2}, \tag{5.37}$$

where

$$\omega_B^{(n)} = \frac{n\pi c}{d \sin \theta_B}, \quad n = 1, 2, 3, \ldots \tag{5.38}$$

is the usual Bragg expression relating the radiation frequency ω_B to the Bragg angle θ_B and the spacing d between crystal planes for a given reflection. The angles θ_x and θ_y are measured relative to the direction in which X-rays incident at the Bragg angle would be diffracted by the crystal, with θ_x measured in the plane containing the incident and reflected direction and θ_y

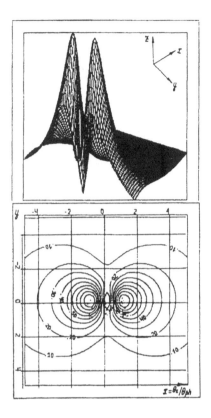

Figure 5.10. The universal angular distribution and the lines of the uniform intensity of the radiation in the PX reflexes; x-axis is in the plane of the vectors $\vec{v}$ and $\vec{\tau}$; z-axis is directed along the vector $\vec{K}_B$; $\theta_B = 30°$ (from [5.11]).

perpendicular to this plane. The crystal thickness is L and the X-ray absorption length is L_a. The definition of θ_{ph}^2 is

$$\theta_{ph}^2 = \frac{m^2 c^4}{E^2} + \theta_s^2 + |g_0|,\tag{5.39}$$

where m is the electron mass, E is the incident electron's kinetic energy, and θ_s^2 represents a mean multiple scattering angle. The Fourier components of the dielectric susceptibility appear as g_0 and g_τ above, where $g_0(\omega_B)$ is the mean value and $g_\tau(\omega_B)$ is the value of the Fourier component corresponding to a reciprocal lattice vector $\vec{\tau}$. This is just the structure factor for the Bragg reflection: $\vec{K} = \vec{K}_0 + \vec{\tau}$, relating the incident wave vector $\vec{K}_0$ to the reflected wave vector $\vec{K}$ by addition of the reciprocal lattice $R\vec{\tau}$ ($|\tau| = 2\pi/d$), which is perpendicular to the crystal planes causing the reflection. For an incident virtual photon, we also have the relation $\vec{K}_0 V = \omega_B$.

The frequency distribution of PX has a narrow bandwidth and the distribution is given by

$$\frac{\partial N}{\partial U} = N_1 J_1(U),\tag{5.40}$$

where

$$J_1(U) = \frac{1 + U^2(1 + \cos^2 2\theta_B)}{(1 + U^2)^{3/2}}\ ,\quad U = \frac{\sin\theta_B}{\cos\theta_B}\frac{\omega - \omega_B}{\omega_B \theta_{\rm ph}}.\tag{5.41}$$

The factor $N_1 = (\pi/2)N_0$, where N_0 is defined to be all the factors in Eq. (5.37) which are independent of θ_x and θ_y, i.e., everything but the last term in Eq. (5.37) for a given value of n in the sum. From Eq. (5.41) it can be seen that the line width will depend on θ_B and ω_B as well as on the electron beam energy, the multiple scattering, and the mean dielectric susceptibility through $\theta_{\rm ph}$ (Eq. (5.39)).

PX angular and spectral distributions have the same form for different crystals because this radiation is generated by a charged particle moving with a uniform velocity. As a result, PX characteristics do not depend on the interaction between the particle and the atoms of the crystal.

5.5 Channeling Effect and Channeling Radiation

If, for example, a proton beam is directed on a single crystal and the intensity of the outgoing beam to be measured depending on the rotation angle of the crystal, then at some directions which coincide with the directions of crystallography axes the crystal becomes 'transparent' (see Fig. 5.11). This effect is called the *channeling effect*, and it can be *axial* or *planar*. An increase in the intensity of penetrating particles means that energy loss of the particles is decreased, or, to put it another way, the particle trajectory passes at a large distance from the atoms. The particles moving in a single crystal at small angles to its axis undergo mirror reflection from these atomic chains.

To understand qualitatively the nature of this phenomenon consider firstly the motion of particles emerging from one of the atomic sites under a small angle ϑ_0 (see Fig. 5.12). We shall take into account only the influence of the Coulomb field of the

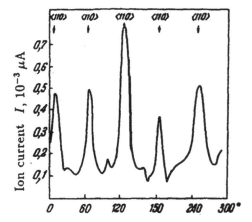

Figure 5.11. The intensity of 75 keV proton beam passing a thin single crystal gold film at different orientations [5.12].

nearest nucleus. If d is the interatomic distance than the impact parameter in relation to the nearest nucleus is $\rho \simeq \vartheta_0 d$. The angle ϑ of the particle deflection will be the same as in Rutherford scattering (see Sec. 5.2) $zZe^2/\rho E = b/\rho$ (we consider the particle as nonrelativistic and its energy is E). As a result the angle of the particle emerging from the atomic row will be

$$\varphi = \theta_0 + \theta = \rho/d + b/\rho. \tag{5.42}$$

It follows from the formula that the minimum angle of emission of the particle from the row φ_{sh} – the angle of shadow – exists. In fact, from the condition

$$\frac{d\varphi}{d\rho} = \frac{1}{d} - \frac{b}{\rho^2} = 0 \quad , \qquad \rho_{\text{min}} = \sqrt{bd} \tag{5.43}$$

it follows that

$$\varphi_{\text{sh}} = 2\sqrt{b/d} = 2\sqrt{E_1/E}, \tag{5.44}$$

where $E_1 = zZe^2/d$.

Thus, for instance, for the atomic row of a tungsten single crystal directed along the axis [100] $E_1 = 340\,\text{eV}$, and therefore 1 MeV-protons emitted from the site of this row are characterized by the angle $\varphi_{\text{sh}} = 2°$. More accurate consideration of the phenomenon leads to approximately the same estimation of the shadow angle.

It follows from the reversibility of mechanical motion that as the particles emitted from atomic row sites cannot emerge under

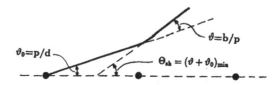

Figure 5.12. The scheme of a scattering of charge particles emerging from the lattice site by the nucleus of a neighbouring atom.

an angle to the axis that is less than φ_{sh}, the particles incident at shallow angles less than φ_{sh} cannot approach this chain either and therefore have to be reflected by the row. If particles come to the row at an angle smaller than the shadow angle, as a rule, the angle of reflection will be large, as trajectories of such particles lie near atomic nuclei. As the angle of incidence decreases it becomes equal to the angle of reflection. The angle of incidence of the particle to the row when the equality of incident and reflected angles begins to hold is called the *angle of channeling*, φ_{ch}. The channeling angle is approximately one and half times less than the shadow angle. A schematic illustration of axial and planar channeling of energetic particles is presented in Fig. 5.13.

One can see from the formula (5.44) that for high-energy particles the probability of channeling is negligible. It is necessary also to note that the charge and the mass of the particle do not explicitly enter into the formula (5.44) for the shadow angle φ_{sh}. The charge of the particle as well as the medium charge influence only the barrier energy E_1 which separates the given channel from the neighboring one.

So, the directions along the crystalline axes and planes are closed for particles emerging from lattice sites. Therefore, if as a result of any nuclear processes (say, α-decay, elastic and inelastic scattering of protons) the sites of a single crystal become particle emitters, then in the directions of axes and planes particular shadows should be observed. This phenomenon was predicted and discovered by Tulinov and is named the *effect of shadows*.

The classical picture of channeling has the particle reflecting back and forth off the boundaries of the channel while propagating down the crystal. For nonrelativistic electrons the radiation associated with this motion is ultraviolet in a vacuum and is rapidly absorbed. For relativistic electrons and positrons the radiation is Doppler shifted to the X-ray region where the material is transparent. From a quantum viewpoint crystalline fields in the direction transverse to the particle's relativistic motion introduce eigenstates,

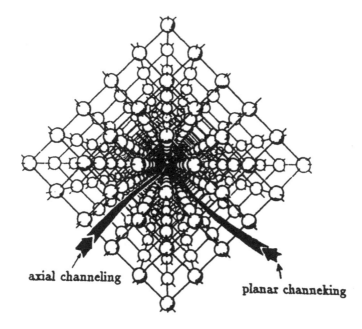

axial channeling

planar channeling

Figure 5.13. Schematic illustration of axial and planar channeling.

and radiation results from spontaneous transitions between these states.

The potentials that steer the incident beam are usually modelled by a so-called continuum approximation, using cylindrically symmetric string potentials for axial channeling and two-dimensional planar potentials for planar channeling. As an example in Fig. 5.14 the Bloch states for 1- and 2-MeV electrons channeled along the $\langle 110 \rangle$ axis of MgO are presented. The wave functions describing the transverse motion are different for the 1- and 2-MeV electrons because of their different effective mass γm. One can see that the difference in transverse energy between the $2p$ and $1s$ states in the Mg ion channel is near 50 eV and so the radiation in the forward direction associated with this transition would be near $50\,eV \times 2\gamma^2 = 2.5\,keV$ (the factor $2\gamma^2$ obeys Doppler shift). We see that the radiation of MeV-electrons will lie in keV-region.

There are a number of reasons for studying radiation from channeled particles: as a method for characterizing the properties of the channeled particles; as a method for characterizing the properties of the crystal; and as a source of radiation.

As well as having the potential for being a relatively narrow

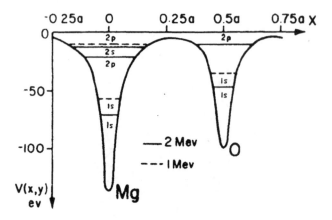

Figure 5.14. Levels of transverse energy for 1- and 2-MeV electrons channeled along ⟨110⟩ axis of MgO. The potential curve is the cross section of the continuum potential $V(x,y)$ along the (001) axis (from [5.13]).

band, the emission has a number of other desirable properties: the photon energy can be tuned by changing the particle energy; if the particle is planar channeled, then the radiation is linearly polarized; and emission occurs in a narrow cone centered about the relativistic velocity direction. In addition, it was first predicted by Kumakhov [5.14] that this radiation can serve as a powerful source of X- and γ-rays.

Most of the work that has been done to date has been concerned with the interactions of H^+ and heavier ions, although there have been a number of experimental studies of β-particle channeling. Electron channeling in particular has been investigated at all energies from a few keV up to several MeV, but the experiments have been carried out with a rather low beam current of the order of nanoamperes. Up to 1990 there was a question whether dechanneling of particles will occur with increasing current due to heat distortion of the crystal lattice and damage. But in 1990 Genz and coworkers [5.15] obtained very promising results on channeling radiation, and this technique is now considered as an intensive source of soft X-rays (see below Chapter 6).

Chapter 6

Electron Accelerators as a Source of Radiation

Electron accelerators are naturally instruments for obtaining high-energy electrons. On the other hand, using the interaction of relativistic electrons with matter we can effectively produce a high flux of gamma quanta, soft X-rays, and neutrons.

Neutrons are secondary particles from electron accelerators, originating in the target as a result of photonuclear reactions (γ, n) and (γ, f). Gamma quanta are produced when fast electrons pass through the matter in the form of bremsstrahlung with continuous spectrum up to $E_{\max}$ equal to the electron kinetic energy. An interaction of relativistic electrons with periodic systems (set of foils or crystalline structures) leads to the origination of quasimonochromatic X-rays due to transition radiation, parametric radiation, and channeling radiation. New possibilities in powerful monochromatic electromagnetic radiation production are opened with the free electron lasers considered in Chapter 4, in which an electron accelerator is the heart of the device. As we will see, the microtron can also serve as a positron beam. All these items, being of great importance for different applications, are considered in detail in this chapter.

6.1 Bremsstrahlung

In an elementary act of electron interaction with the Coulomb field of the target nucleus electromagnetic radiation originates, and this radiative loss is proportional to the number of target nuclei (concentration of the target nuclei). Hence, it is natural to increase the bremsstrahlung power using a thick target, however, this seemingly simple method leads to additional difficulties:

(i) nonradiative losses of electron energy leading to its energy straggling;
(ii) photon absorption in a radiator;
(iii) multiple radiation of photons by the same electron;
(iv) multiple elastic scattering of electrons before radiation transition.

Therefore, analytical calculation of the energy and angular distribution of bremsstrahlung is a very complicated problem, and is performed numerically (see, for instance, [6.1]). On the other hand, there are a lot of experimental characteristics of bremsstrahlung from different kinds of targets that make it possible to design a bremsstrahlung target in accordance with experimental demands.

By 'optimal target' we mean a bremsstrahlung target which provides a maximum forward yield of radiation at given energy of electrons striking the target at the right angle. In the electron energy region $5-30\,\mathrm{MeV}$ for heavy elements an optimal target thickness equals 0.3 rad.length — Fig. 6.1. For instance, the optimal thickness of a tungsten target is equal to $1\,\mathrm{mm}$.

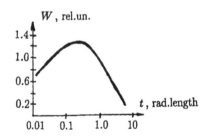

Figure 6.1. Dependence of the forward bremsstrahlung power on target thickness (target material – Au, electron energy – 17 MeV) [6.2].

The forward bremsstrahlung power behaviour with electron energy (Fig. 6.2) is approximated for heavy elements (Ta, W, Au, Pb) by the formula

$$P(E_0) = 82 \cdot E_0^{2.63}. \qquad (6.1)$$

Here $P(E)$ is the power of the absorbed dose in air in rad·m^2 per mA·min, and E_0 is the electron energy in MeV.

The optimal target thickness is considerably less an electron range, that is why light element absorbers (usually aluminium or graphite) are placed behind the target to shield samples, and parts of experimental set-up which are behind the target. The object in using absorbers is to stop fast electrons slightly disturbing γ-radiation. To estimate the necessary thickness of an absorber it

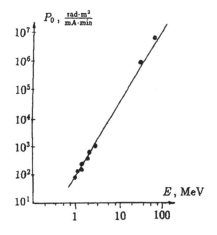

Figure 6.2. Radiation dose of forward bremsstrahlung with electron energy [6.3].

is useful to know that for electron with energy more than 1 MeV the range R_e in Al with sufficient accuracy equals

$$R_e = 0.5\,E, \qquad (6.2)$$

where R_e is in g/cm², E in MeV.

Unlike the Z^2-dependence of the radiation conversion efficiency of electron energy at elementary act, it is changed almost linearly with atomic number in thick targets (Fig. 6.3). It is necessary to note, that this efficiency is rather high, and reaches 30% for heavy elements at the electron energy ~ 10 MeV.

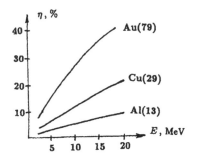

Figure 6.3. Conversion efficiency of electron energy into radiation for Au, Cu and Al [6.5].

Figure 6.4 illustrates the forward spectrum of bremsstrahlung from an optimal tungsten target at electron energies 10, 15 and 30 MeV. These spectra are numerically calculated taking into account all the considerations mentioned above [6.4].

The energy spectrum substantially depends on the observation angle, especially for thick targets. The slowing down of electrons due to ionization losses leads to reduction of the high-energy part in relation to low-energy radiation. The larger the angle, the

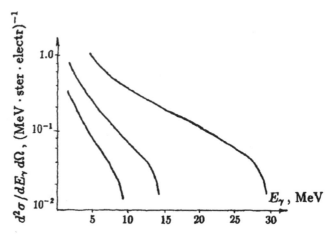

Figure 6.4. Forward bremsstrahlung spectra at different electron energies – 10, 15 and 30 MeV [6.4].

more this transfer into low-energy region. An analogous spectrum transformation occurs due to multiple radiation of photons by a single electron. One can see this spectrum transformation in Fig. 6.5, where bremsstrahlung of 10 MeV electrons at the angles 0° and 10° are given.

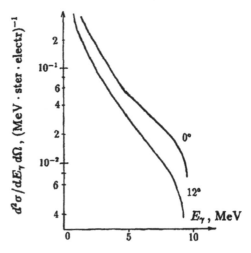

Figure 6.5. Bremsstrahlung spectrum of 10 MeV electrons at different angles [6.4]

It is natural that angular distributions of different energy regions of bremsstrahlung are different: the high-energy part is concentrated in narrower angle interval than the low-energy part.

This statement is illustrated by Fig. 6.6, where experimental data on half-width of angular distribution of the whole spectrum and its high-energy part $(E_\gamma \geq 5\,\mathrm{MeV})$ are given with electron energy.

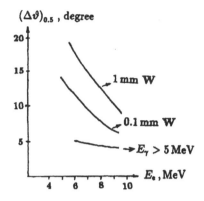

$(\Delta\vartheta)_{0.5}$, degree

1 mm W

0.1 mm W

$E_\gamma > 5\,\mathrm{MeV}$

E_e , MeV

Figure 6.6. Half-width of bremsstrahlung energy distribution at different electron energies and for different tungsten target thickness [6.6].

The formula of bremsstrahlung angular distribution can be found in the same way as Lawson's calculation [6.7]. In experiments we usually have to take into account not only intrinsic divergence of the bremsstrahlung from the target but also the divergence of the electron beam originating during acceleration and due to passing through the extracting window.

Let the intrinsic bremsstrahlung angular distribution at the elementary act of scattering follow Gaussian

$$H(\theta) = \frac{E^2}{2\pi\mu^2} \exp\left(-\frac{\theta^2 E^2}{2\mu^2}\right) , \qquad (6.3)$$

where E is the electron kinetic energy, and $\mu = mc^2 = 0.511\,\mathrm{MeV}$. The root-mean square scattering angle for an electron in its passage through the extracting aluminium foil of thickness t_{Al} (in radiative length) is determined according to Rossi and Greisen [6.8] as

$$\overline{\theta^2} = \frac{440 t_{\mathrm{Al}}}{E^2} , \qquad (6.4)$$

and angular distribution of electrons after passing the radiator of thickness t_r is also Gaussian

$$F(t, \theta) = \frac{E^2}{440\pi t_\mathrm{r}} \exp\left(-\frac{E^2\theta^2}{440 t_\mathrm{r}}\right) . \qquad (6.5)$$

As for the radial divergence of the electron beam, we can also consider it follows the normal distribution

$$G(\theta) = \frac{1}{2\pi\theta_e^2} \exp\left(-\frac{\theta^2}{2\theta_e^2}\right), \tag{6.6}$$

with the root-mean square angle θ_e^2. All distributions are normalized in such a way that the integral of the distribution function over all angles equals 1.

Considering bremsstrahlung angular distribution from a thick target we shall neglect ionization losses, absorption of radiation in the target is insignificant, and every electron emits only one photon, i.e., the bremsstrahlung angular distribution is mainly formed by elastic scattering of electrons in the target material.

The angular distribution of radiation from a layer of thickness δt placed in a depth t is the convolution of four Gaussians (6.3–6.6), i.e., the radiation in a unit solid angle from a layer δt is

$$\delta R = \sigma\delta t \frac{E^2}{2\pi\mu^2} \frac{E^2}{440\pi t_r} \frac{E^2}{440\pi t_{Al}} \frac{1}{2\pi\theta_e^2} \int\int\int d^3\varphi \sin\theta''' \times$$

$$\sin\theta'' \sin\theta' d\theta''' d\theta'' d\theta' \exp\left(-\frac{\theta^2 E^2}{2\mu^2}\right) \exp\left(-\frac{E^2(\theta'-\theta)^2}{440 t_r}\right) \times$$

$$\exp\left(-\frac{E^2(\theta''-\theta')^2}{440 t_{Al}}\right) \exp\left(-\frac{(\theta'''-\theta'')^2}{2\theta_e^2}\right), \tag{6.7}$$

where $\sigma\delta t$ is a fraction of electron energy transformed into radiation in the bremsstrahlung target of thickness δt.

It is mathematically easier to calculate expression (6.7) in the following way. It is well known that a sum of any number of independent normal distributions is also normal, and the dispersion of the sum of independent random quantities equals

$$\sigma^2\left[\sum_i \xi_i\right] = \sum_i \sigma^2(\xi_i). \tag{6.8}$$

Therefore, we can straight away write that the distribution δr has the form

$$\delta R = \text{const} \cdot \exp\left[-E^2\theta^2/440(t_r + t_{Al}) + 2\mu^2 + 2\theta_e^2 E^2\right]. \tag{6.9}$$

The value of the constant is simply to calculate from the condition that the distribution has to be normalized:

$$\int\limits_0^{2\pi}\int\limits_0^{\pi} \delta R \sin\theta \, d\theta \, d\varphi = 1. \tag{6.10}$$

In our approach $\sin\theta \simeq \theta$, and integrating over θ from 0 to ∞ we obtain

$$\delta R = \frac{E^2 \sigma \delta t}{2\pi[220(t_r + t_{Al}) + \mu^2 + \theta_e^2 E^2]} \times$$

$$\exp\left[-\frac{E^2\theta^2}{2(\mu^2 + 220t_r + 220t_{Al} + \theta_e^2 E^2)}\right]. \qquad (6.11)$$

By integrating of Eq.(6.11) over the total target thickness from 0 to t_r we have

$$R(\theta) = \frac{\sigma E^2}{440\pi}\left[-\text{Ei}\left\{-\frac{E^2\theta^2}{440(t_r + t_{Al} + 2\mu^2 + 2\theta_e^2 E^2)}\right\} + \right.$$

$$\left. \text{Ei}\left\{-\frac{E^2\theta^2}{2(\mu^2 + 220t_{Al} + \theta_e^2 E^2)}\right\}\right], \qquad (6.12)$$

where Ei is the integral exponential Euler function

$$-\text{Ei}(-x) = \int_x^\infty \frac{e^{-x}}{x}\, dx. \qquad (6.13)$$

The bremsstrahlung angular distribution obtained for a thick target describes well enough the experimental data [6.5]. It is necessary to underline that the estimation of angular divergence of radiation $\vartheta \simeq mc^2/E$ is still valid for γ-quanta of maximum energy.

Consider the following problem: is it possible to form a radiation field with reflectors? The problem is reduced to determination of an albedo of γ-radiation, i.e., to characteristics of backscattered γ-radiation, falling on a semi-infinite medium. Theoretical calculations of such processes are performed by the Monte-Carlo technique [6.7], and experiments with bremsstrahlung were made by Zhalsaraev et al. [6.9].

The effects observed are mainly the Compton effect and pair production leading to γ-ray scattering onto large angles. So, it is clear, that the energy spectrum of backscattering radiation has to be much softer than the incident one, and it has a maximum close to $mc^2/2 = 0.255\,\text{MeV}$ (Compton effect at $180°$) and $0.511\,\text{MeV}$ (positron annihilation). In a soft range substantial contributions from characteristic röntgen radiation and bremsstrahlung from secondary electrons take place. As an example, the differential on the energy and angle of the bremsstrahlung albedo at the angle $25°$ for $E_{\text{max}}{=}6\,\text{MeV}$ is shown in Fig. 6.7.

The dependence of the bremsstrahlung albedo a on the atomic number of the target is rather weak, and only slightly increases

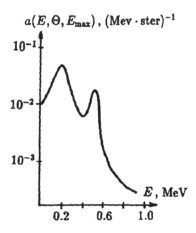

Figure 6.7. Albedo of bremsstrahlung at $E_{max} = 6$ MeV — the number of quanta in a unit energy interval backscattered into unit solid angle in the direction of 25° per one quantum of bremsstrahlung for semi-infinite stannum target (Z=50) (from [6.9]).

in the region of small Z due to an increase in the contribution from the Compton effect as seen in Fig. 6.8).

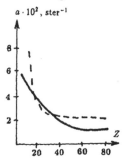

Figure 6.8. Z-dependence of the bremsstrahlung albedo: solid line — experiment, dashed line — calculation (from [6.9]).

Therefore, we see that reflectors cannot lead to increasing the yield of photonuclear reactions, but they can play useful role in detecting secondary radiation.

6.2 Photoneutrons

Generation of neutrons at electron accelerators occurs through a (γ, n)-reaction. Materials with large Z are more preferable for the problem as they have the lowest neutron binding energy (with the exception of deuterium and beryllium that have abnormally low neutron binding energy; we shall consider these elements later). Figure 6.9 shows the neutron yield behavior with electron energy for lead, plutonium and uranium. In the case of fissionable nuclei the yield is approximately twice as large due to fission neutrons. However, the use of U or Pu, because of the high specific radioactivity of fission fragments, demands special precautions, and leads to some difficulties in their exploitation. Tantalum and tungsten give approximately the same neutron yield as lead. To generate neutrons a target-convertor is directly bombarded by electrons or bremsstrahlung from a closely placed target.

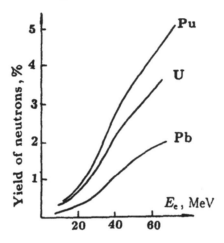

Figure 6.9. Energy behaviour of neutron yield for Pb, U and Pu [6.10].

A photoneutron spectrum (Fig. 6.10) is similar to the spectrum of fission neutrons, it has a maximum at neutron energy of about 1 MeV and isotropic angular distribution, which means that this reaction passes over a stage of compound nucleus formation.

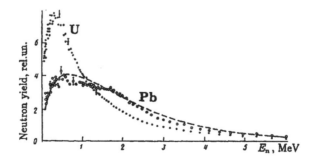

Figure 6.10. Photoneutron spectra from Pb and U at E_e=45 MeV [6.11].

Considering an electron accelerator as a neutron source, it is necessary to note that using different converters it is possible to form different fast neutron spectra and to use reactions of the type $(n, n'), (n, p), (n, \alpha)$ in different applications. We will now turn our attention to the nuclei with low neutron binding energies — beryllium and deuterium. The Be and D nuclides possess low charge Z and so have small level density, and actually the spectrum of neutrons originated in these nuclei, reflects the spectrum specificity of their low-lying states. Figure 6.11 shows experimentally measured photoneutron spectra originated in Pb-, Be-, and D_2O-converters [6.12]. It is seen that every converter possesses its own characteristic fast neutron spectra.

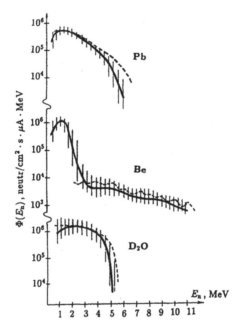

Figure 6.11. Fast neutron spectra from Pb-,Be- and D_2O-converter at $E_{max} = 15\,MeV$; solid lines — experiment [6.12], dashed lines — [6.13, 6.14, 6.15].

As Be and D have very low neutron binding energies, the neutron yield per electron in these nuclei is larger than in heavy ones [6.16], and it seems that a higher thermal neutron flux could be obtained. Nevertheless, the highest neutron flux density is obtained by using heavy element converters, and this is explained by the fact that the (γ, n) cross section for light nuclei is lower due to the smaller nuclear radii. Besides, the γ-quanta absorption cross section for heavy nuclei is higher than for light ones, i.e., their range is less in Pb than in Be. Therefore a light nuclei converter is more lengthy (see Fig. 6.12) than, for example, Pb-converter which is more similar to an ideal point source. This point should be borne in mind when designing the neutron converter.

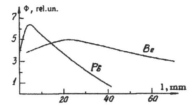

Figure 6.12. Fast neutron distribution along the element of a Pb- and Be-converter cylinder.

The inevitable presence of γ-quanta in an electron accelerator based neutron source may of course be a cause of the background for neutron experiments. The natural way out of this problem is to decrease the electron energy. It is very easy to do so for electron accelerators, and in such a case, particularly in a microtron, the beam power practically remains constant with the energy change. Therefore, although the energy dependence of the neutron yield is rather strong, the neutron yield under the constant beam power has a smooth energy dependence. Thus, for example, while changing an electron energy from 30 MeV to 15 MeV the neutron yield decreases only by 1.5 times (see Fig. 6.13).

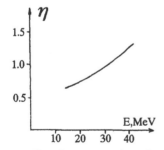

Figure 6.13. Neutron yield (in relative units) with electron energy of a constant beam power ($P = EI$ =const); neutron yield at 30 MeV equals 1.

The last question is how much we can now increase the neutron yield by increasing the average power of the accelerating beam? We will consider this problem for a microtron. The modern electronic industry can supply physicists with a new powerful high-frequency source — klystron amplifier with pulse power 4-5 MW and an average power of 20–25 kW. This is approximately 10 times higher than magnetrons with average power of ∼ 2 kW, which are widely used in. With such a source it is quite feasible to obtain an electron beam with an average power of ∼ 10 kW.

True, the problem of cooling the target or other accelerator details arises, but it is a completely solvable problem for modern engineering. In such a case one can obtain a thermal neutron flux $\sim 10^{10}$ n/(cm^{2}s) from a Pb-converter. By using fissionable elements this value can be increased twice. This is still rather a high flux.

It is basically possible to use neutron multiplication, say, with 10-, or even 100-fold increase in the amount of neutrons [6.17]. An accelerator with such a set-up becomes comparable (in the sense of neutron flux value) with a standard reactor. Of course, multiplication targets demand more careful radiation control, and they will be available only to large laboratories or enterprises.

6.3 X-ray Production by Relativistic Electrons

In chapter 5 we considered the main principles of interaction of electrons with inhomogeneous media that leads to the origination of transition, parametric, or channeling radiation depending on the properties of the media used. Now we turn our attention to the experimental results obtained, and compare the results with theoretical predictions on intensity, angular and spectral distribution of radiations. Special attention will be paid to considering the experimental set-up for producing X-rays by relativistic electrons.

6.3.1 X-ray Generation from Transition Radiation

X-rays are produced by transition radiation when high-energy electrons cross the interface between two media or between a vacuum and a medium. The frequency range of the transition radiator is governed by the desired application of the source. The peak and width of the spectrum are optimized by the selection of the foil thickness and number. In addition, we can select a radiator material with a particular photoabsorption edge which will truncate the spectrum above the photoabsorption-edge photon energy. The peak of the radiator spectrum can be adjusted by maximizing the single-foil resonance term, $4\sin^2(l_2/Z_2)$. This is accomplished by designing the foil thickness to be resonant at the desired frequency, ω_d, by setting $l_2 = (\pi/2)Z_2(\omega_d)$, or:

$$l_2 \simeq \frac{\lambda}{\left(\frac{2}{\gamma^2}\right) + \left(\frac{\omega_2}{\omega_d}\right)^2}. \tag{6.14}$$

For no phase addition between foils, the foil spacing, l_2, needs only to be much greater than its formation zone: $l_1 \gg z_1$. Since

z_1 is usually quite small for moderate electron beam energies, the spacing is determined by the minimum required thickness of the spacer for rigid support of the foil.

A minimum electron-beam energy is required to achieve adequate photon production at ω_d. From the requirement that $\omega_d < \gamma\omega_2$, and given plasma frequency ω_2, the minimum electron-beam energy for photon production is obtained:

$$E > \frac{E_0\omega_d}{\omega_2}. \qquad (6.15)$$

To be in the range where the photon production is roughly linear with the electron beam power, an electron-beam energy roughly two times larger than the minimum energy is selected. This selection is dependent on the cost of the construction of the source or on the availability of an existing source's maximum electron-beam energy.

From the discussion of Eq. (3.30) the number of foils is given by

$$N = \frac{2}{\sigma(\omega_d)} = \frac{2}{\mu(\omega_d)l_2(\omega_d)}, \qquad (6.16)$$

where we have again set $\omega = \omega_d$, and the mass-absorption coefficients $\mu(\omega_d)$ one can take from the tables.

Given an adequate electron-beam energy, we can move the peak of the X-ray spectrum around the photon energy using the single-foil resonance term and setting the foil thickness to be as given by Eq. (6.14). This is demonstrated in Fig. 6.14 where a beryllium radiator is designed to be single-foil resonant at various photon energies from 1 keV to 3.5 keV.

As to the spatial distribution of transition radiation, it is emitted, like Cherenkov radiation, in an annular cone (see Fig. 6.15). The cone angle θ is approximately $1/\gamma$, with a width that also is about $1/\gamma$. A large amount of the X-ray power is also at cone angles greater than $\theta = 1/\gamma$. Although the power density drops dramatically after $2/\gamma$, there is radiation emitted out to

$$\theta_{\max} = \sqrt{\frac{1}{\gamma^2} + \left(\frac{\omega_2}{\omega}\right)^2}. \qquad (6.17)$$

For large γ, then $\theta_{\max} \simeq \omega_2/\omega$. For beryllium foils $\omega_2 = 26.1$ eV, for $\omega = 1500$ eV, $\theta = 18$ mr.

Figure 6.15 shows the image of transition radiation obtained by a 150 MeV electrons beam in their passage through 10 foils of aluminium.

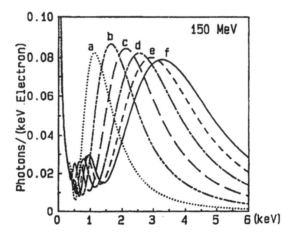

Figure 6.14. The calculated spectral photon density as a function of X-ray photon energy for beryllium foil radiators optimized to radiate at 1.0, 1.5, 2.0, 2.5, 3.0, and 3.5 keV. The peak of the spectrum can be moved by designing the single-foil resonance term to be maximum at the desired photon energy. The foil stack parameters are as follows: a) 11 foils 1.76 μm thick, b) 26 foils 2.53 μm thick, c) 48 foils 3.2 μm thick, d) 81 foils 3.75 mum thick, e) 126 foils 4.18 μm thick, and f) 187 foils 4.49 μm thick (after [6.18]).

At first it would appear that the higher the electron-beam energy, the higher the power and the smaller the image spot size, and, hence, the higher the power density. However, as we have just seen, much of the X-ray flux is larger than $1/\gamma$, going out to $\theta \simeq \omega_2/\omega$ regardless of electron-beam energy. Thus for the most efficient X-ray production, X-ray collection optics is required.

By designing transition radiators to emit X-rays at the foil material's K-, L-, or M-shell photoabsorption edge energies, the bandwidth is narrower than that of radiators. Indeed, the bandwidth of the photoabsorption-edge transition radiator can be reduced by a factor of 2 or more when compared to that of a transition radiator which is not designed at the photoabsorption-edge frequency.

Since the X-ray absorption is reduced on the low frequency side of the photoabsorption-edge, the number of foils that one can use can be large, permitting intense X-ray production. As an example of this effect at soft-X-ray photon energies, we compare in Fig. 6.16 the frequency spectra of two transition radiators, one composed of aluminium foils and the other, of beryllium. The aluminium radiator spectrum is dramatically truncated above

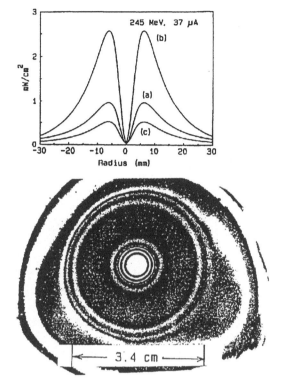

Figure 6.15. *Top:* The calculated power density as a function of beam radius for transition radiators composed of (a) 25 foils of 1.0-μm beryllium, (b) 25 foils of 2.1-μm beryllium, (c) 12 foils of 1.0-μm aluminium; the distance between the radiator and detector is 3 m; the electron beam energy is 245 MeV and current is 37 μA. *Bottom:* Exposure of Shipley MICROPOSIT ECX-1029 photoresist on 10-cm diameter silicon wafer using soft X-rays from a transition radiator. The exposure time was 150 s. The annulus of the transition radiation cone is imaged in the resist. The electron-beam current was 35 mA; the beam energy was 150 MeV. The radiator was 10 foils of 1.0 μm aluminium (after [6.18]).

1.559 keV while the beryllium's is not. The beryllium stack's K-edge is at 112 eV and does not affect the spectrum. The absorption of the X-rays in the aluminium foils increases by a factor of 10 as the X-ray photon energy passes the K-edge. This results in X-rays with photon energies above 1.559 keV being absorbed in the foils, abruptly truncating the frequency spectrum.

Figure 6.17 illustrates experimental results on soft X-ray generation by low-density foils of beryllium, aluminium and magnesium [6.19]. The number of photons per unit frequency per unit sol-

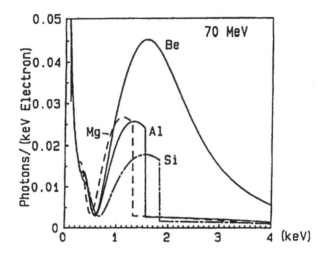

Figure 6.16. The calculated spectral photon density as a function of X-ray photon energy for various optimized foil stacks. The electron-beam energy is 70 MeV. The foil stack parameters are as follows: 61 foils of 1.84 μm beryllium; 14 foils of 1.19 μm aluminium; 17 foils of 1.44 μm silicon; and 14 foils of 1.53 μm magnesium. The figure shows the brightness of the beryllium-foil stack compared to the other stacks and the effect of the K-shell photoabsorption edge of the spectra on the magnesium, aluminium and silicon radiators (after [6.18]).

id angle per electron is plotted as a function of photon energy. The spectra were produced from 25 and 54 MeV electrons. The aluminium and magnesium spectra were highly influenced by the respective K-shell absorption edges. The sharp fall-off of the spectra has been somewhat masked by the bandwidth of the detector. However, the bandwidth narrowing character of the K-edge effect and shift in mean frequency between the two spectra are clearly visible.

Even without the effect of the photoabsorption edges, hard X-rays can be generated efficiently by using foils of high-density materials such as gold, stainless steel, or copper. It is important in this case to select foil materials with thickness and densities that minimize the bremsstrahlung and maximize the transition radiation. For example, iron (stainless steel) and copper foils are excellent candidates since they have comparatively high densities and moderate atomic numbers. High-density foils which also have high atomic number such as gold or tungsten can be used if it is desirable to lower the electron-beam energy further and if extremely hard bremsstrahlung contamination of the transition radiation spectrum does not matter.

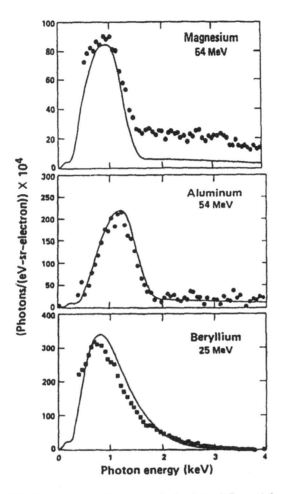

Figure 6.17. Predicted and measured absolute differential production efficiency of the difference spectrum for the transition radiation emitted for 18 foils of 1 μm beryllium, 30 foils of 1 μm magnesium, and 30 foils of 1 μm aluminium (from ref.[4.20]). The magnesium spectrum is truncated at the 1.303 keV K-shell photoabsorption edge, and the aluminium spectrum is truncated at the 1.559 keV K-shell photoabsorption edge. The truncation of the spectrum is somewhat blurred by the detector frequency resolution. The calculated spectra include the effect of the detector resolution (bandwidth).

To illustrate this statement Fig. 6.18 shows the result of measuring the transition radiation from a gold radiator made by

Piestrup and coworkers [6.21]. The target consisted of 10 foils of 1 μm-thick gold, the energy of electron beam was 105 MeV. A single 10 μm foil of gold was used to account for the bremsstrahlung and other spurious ionizing radiation from the accelerator. By subtracting the single-foil data from the foil-stack data and obtaining a difference spectrum, these other source of radiation was eliminated.

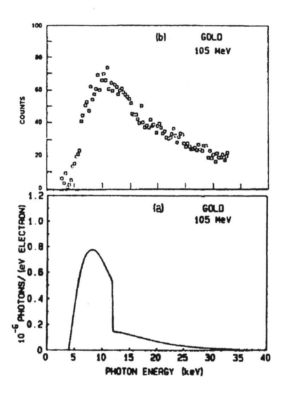

Figure 6.18. Predicted (a) and measured (b) spectral photon density of the difference spectrum for 10 foils of 1 μm gold (after [6.20]).

It was also shown [6.21] that using photoabsorption-edged transition radiation one can obtain a quasimonochromatic X-ray source. A quasimonochromatic source was designed using a titanium foil stack. Titanium is of moderate density (4.5 g/cm²) and has a K-shell photoabsorption edge at 5.0 keV.

The target consisted of 10 foils of 2 μm titanium. A single 28 μm-titanium foil was used to determine the bremsstrahlung and spurious radiation background. The measured radiation generated by the single 28 μm foil was subtracted from that produced by the foil stack. The spectra from the 28 μm foil and 10-foil transition radiator are shown in Fig. 6.19.

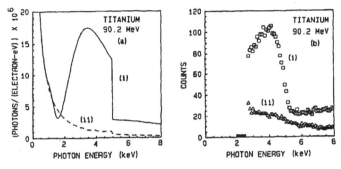

Figure 6.19. Predicted (a) and measured (b) spectral photon density for 10 foils of 2 μm titanium and an equivalent thickness single foil of titanium. The electron-beam energy was 90.2 MeV. The emission from a single 28 μm foil is also shown (ii) (after [6.21]).

6.3.2 Experimental Study of Parametric Radiation

Experiments investigating parametrically generated X-rays were performed in 1991 with a silicon crystal, using a 90 MeV-linac [6.22]. A schematic diagram of this PX experimental set-up is shown in Fig. 6.20.

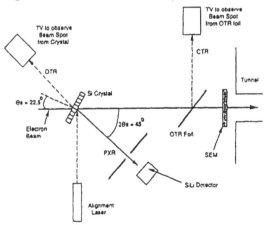

Figure 6.20. Apparatus for parametric X-ray experiment. The electron beam enters from the left, striking the silicon crystal. The X-rays emerge at the Bragg angle and are detected by a Si(Li) detector. The electrons continue on to an OTR foil monitor station and on SEM (from [6.22]).

The figure shows a silicon crystal oriented for production of PX from the ⟨011⟩ planes. The electron beam is incident on these planes at a Bragg angle of 22.5°, and the detection angle relative to the electron beam direction is 45°. In order to successfully generate PX at a given Bragg angle, the direction of the electron beam relative to the diffracting planes must be maintained within a tolerance of a few milliradians. In order to accomplish this, the experimental arrangement makes use of optical transition radiation diagnostics (OTR) (see [6.23–6.25]) to monitor the electron beam and its direction with respect to the crystal planes. Both the OTR and PX are observed at large angles relative to the beam direction, which significantly simplifies the tuning of the system.

A detailed drawing showing the silicon crystal's orientation is given in Fig. 6.21a. The crystal is cut so that ⟨011⟩ planes are oriented at right angles with respect to the crystal face, whose normal is the ⟨100⟩ direction. A square 4 cm² area centered on the beam axis is etched out of the crystal to reduce its thickness to 20 microns.

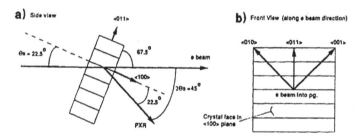

Figure 6.21. *a)* Detailed view of crystal orientation relative to the electron beam for PX generation from the ⟨011⟩ planes of silicon at a Bragg angle of 22.5 °. *b)* Front view of the silicon crystal showing the boundaries of the area of the crystal that was etched to 20 μm thickness (from [6.22]).

Figure 6.21b shows the boundaries of the etching with respect to the crystal lattice vectors. The side of the crystal away from the electron beam is an optically mirrored surface. The etched surface (nearest the electron beam) is also highly reflective in the visible. The crystal is initially oriented so that the ⟨011⟩ is aligned to the vertical (normal to the plane of incidence). The Si(Li) detector was positioned one meter away from the crystal's vertical rotation axis and was used to obtain the energy spectrum of the PX radiation.

Spectra taken at the Bragg angles of 22.5° for PX generation from the ⟨011⟩, and ⟨111⟩ planes are shown in Fig. 6.22. Note that the $n = 2$ mode is the first mode allowed for PX generation from the $\langle 011 \rangle$ plane. The Bragg condition predicts the energies for an incidence angle of 22.5° are 8.42 keV and 5.16 keV respectively for the $\langle 011 \rangle$ and $\langle 111 \rangle$ planes.

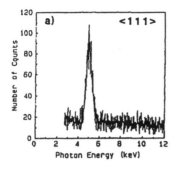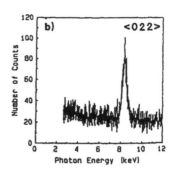

Figure 6.22. *a)* Measured spectrum of X-rays from the $\langle 111 \rangle$ reflection of silicon at Bragg angle of 22.5 °, with peak at 4.96 keV. *b)* The spectrum from the $\langle 022 \rangle$ reflection at Bragg angle 22.5 °, with a peak at 8.44 keV [6.22].

It was demonstrated in the experiment that the observed radiation possessed the energy, line width, and angular pattern predicted by the theory of parametric X-ray generation. Furthermore, this radiation was produced with a linac of modest beam quality, and moderate energy, i.e., a divergence of about 1 milliradian and an energy of 90 MeV.

6.3.3 Channeling Radiation

In 1990 Genz *et al.* [6.26] obtained very promising results on channeling radiation using the continuous wave low emittance electron beam ($\epsilon = 0.04 \, \pi \text{mm·mrad}$) from the superconducting injector of the Darmstadt electron accelerator. The achieved beam divergence was of $\theta = 0.05 \, \text{mrad}$, and it is established that the beam divergence is two orders of magnitude less than the value of the critical angle in channeling, which amounts typically to 5 mrad for the 10 MeV electron-bombarding energy. Thin silicon and diamond crystals were studied for electron impact energies between 3 and 8 MeV. A diamond crystal was chosen since it has been established before that its channeling radiation spectrum is governed merely by a few, energetically close-lying transitions. These make this crystal quite an interesting candidate for a narrow bandwidth photon source. Furthermore, due to the low Z material, disturbing radiation caused by bremsstrahlung is less important and the large value of thermal conductivity as well as the unusual large

Debye temperature (1850 K) of diamond suggest that crystal may well resist the needed high beam currents for producing a really intense source.

The channeling radiation spectra were detected at $0°$ with respect to the beam axis. In addition, the authors employed a 10-μm-thick silicon crystal aligned along the $\langle 111 \rangle$ axis, and a 50-μm-thick diamond of 4 mm diameter aligned with respect to the $\langle 110 \rangle$ axis was investigated. Beam currents were varied between 3 nA and 30 μA. For electron currents of 3 nA a Si(Li) detector was used, which was connected directly to the vacuum of the scattering chamber and shielded against background radiation by 60 cm of lead. For the high current experiment LiF thermoluminiscence dosimeters were used for the detection of channeling radiation.

Examples of axial channeling radiation spectra obtained by bombarding the two crystals described above with electrons of 5.3 and 6.7 MeV respectively, are shown in Fig. 6.23 in the upper part of the figure for diamond (left side) and silicon (right side). The spectra are background corrected, i.e., contributions caused by bremsstrahlung, which are easily detected by tilting the crystal out of axis, are subtracted. The ratio of channeling radiation to bremsstrahlung amounts to about 5:1 in the case of diamond and 1:1 for silicon. The spectra were taken at a current of 3 and 10 nA for diamond and silicon respectively, and it took only 240 seconds to collect the intensity displayed in the figure. A comparison between the two spectra shows that diamond emits four lines at the selected electron impact energy and silicon emits five. Those lines have been identified as being due to various transitions between bound states in the crystal potential.

By increasing the beam current successively from 3 nA up to 30 μA, photon intensities for nine different current settings were recorded. The result is displaced in the form of open circles in a double logarithmic representation in Fig. 6.24. The particular electron energy employed was $E_0 = 5.4$ MeV. For the highest electron current applied, i.e., $I = 30 \mu$A, a photon intensity of $2 \cdot 10^{10}$ photons/s within an energy window of $\Delta E/E \simeq 0.1$ was produced. This is six orders of magnitude more than ever observed before.

During these studies the diamond crystal was kept at room temperature by means of Peltier cooling. The total number of electrons that have passed through the crystal in the course of the investigations amounts to 10^{18} electrons/mm². After this high electron current treatment, a further channeling spectrum was detected by means of the Si(Li) detector at 3 nA. Within the

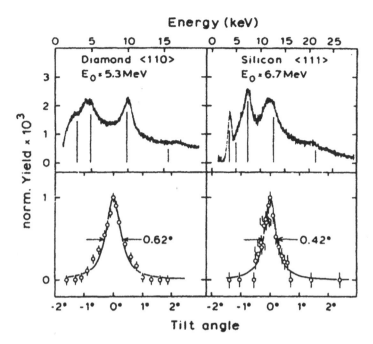

Figure 6.23. Typical channeling spectra obtained by bombarding a diamond crystal along the $\langle 110 \rangle$ direction and a silicon crystal along the $\langle 111 \rangle$ direction with electrons of 5.3 and 6.7 MeV respectively (upper part). The solid lines correspond to transitions as predicted by the single string calculations. Tilting curves obtained by moving the crystal out of axis (lower part). The plotted intensity contains only the $2p - 1s$ transitions [6.26].

statistical uncertainty no difference of the spectra before and after electron bombardment of 10^{18} electrons was observed, which seems to indicate that dislocations have apparently not been produced by the current applied so far, or if they have been produced they annealed themselves during bombardment.

Thus, intense and tuneable photon sources of 10^{10} photons/s and probably much higher can be realized in the X-ray energy region also above 10 keV and most likely also for 50-100 keV (note that the X-ray energy in channeling radiation changes as $\gamma^{3/2}$). The intensities achieved are already sufficient for recording X-ray images in medical and material research applications, for example, with film screen systems or storage phosphor screens.

For comparison, let us describe the parameters of the synchrotron radiation from VEPP-3 (Novosibirsk). The photon flux

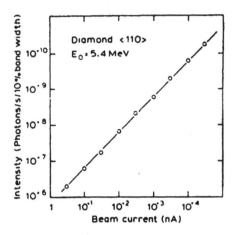

Figure 6.24. Photon intensity for the $2p - 1s$ transition at about $10\,\mathrm{keV}$ as a function of the electron beam current [6.26].

in this machine at energy $2.1\,\mathrm{GeV}$ and the current $30\,\mathrm{mA}$ at photon energy $10\,\mathrm{keV}$ is equal in the energy interval $\Delta E/E = 1\%$ to $4 \cdot 10^{12}$ photons/s·mrad from the turning magnet, and $2 \cdot 10^{14}$ from the superconducting wiggler. As the radiation power is proportional to γ^4 and the maximum of radiation is shifted with energy as γ^3, we see that the brightness of channeling radiation may be the same as from $500\,\mathrm{MeV}$ storage ring.

6.4 A tuneable monochromatic gamma-ray source

Intense, energetic, and monochromatic gamma-ray sources can provide an important tool for a wide field of applications ranging from nuclear to solid-state physics. Up to now the ideal, intense, tuneable monochromatic gamma-ray source with high energy resolution and low divergence does not exist. Instead, different kinds of gamma-ray sources, each with their advantages and limitations, are used. Radioactive decay provides gamma rays with very well defined energies but which are not tuneable. Bremsstrahlung from decelerated charged particles gives an intense gamma-ray spectrum over a very wide energy range. Synchrotron radiation facilities deliver a very intensive and focused X-ray beam, but with a rather large energy spread and low energy ($E_\gamma < 150\,\mathrm{keV}$). More or less tuneable gamma-ray sources have been constructed using Compton scattering, tagged photons, laser backscattering or using synchrotron radiation. All three have rather low luminosity and energy resolution. Synchrotron radiation is clearly the most powerful method for relatively low energies but requires a large infrastructure.

Recently a new concept of obtaining a tuneable monochromatic gamma-ray source was put forward by Jolie and was realized at the Ghent 15 MeV linear electron accelerator [6.27]. A schematic layout of the tuneable γ-ray source is shown in Fig. 6.25.

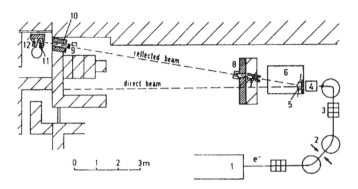

Figure 6.25. Schematic layout of the tuneable gamma-ray source (from [6.28]).

The γ-rays created in the bremsstrahlung source (1) fall on a bent Si single-crystal (2). The single-crystal is curved such that the crystal planes focus at point (3) at a distance of 10.65 m. The gamma rays are reflected via the Bragg relation

$$2d \sin \theta_B = n\lambda = nhc/E \qquad (6.18)$$

on different points on the Rowland circle (4). Each point on the Rowland circle corresponds to one energy for a given order of diffraction. In (6.18) θ_B is the Bragg angle and d the spacing between the reflecting crystal planes.

Gamma-rays that are incident on the crystal planes at angles θ_B can undergo nth order diffraction if their energy E, and associated wavelength λ, fulfills the Bragg-Laue condition

$$n\lambda = \frac{nhc}{E} = 2d \sin \theta_B \qquad (6.19)$$

in transmission. In Eq. (6.19) n denotes the order of reflection, h Planck's constant, c the velocity of light, and d the lattice spacing. The diffracted gamma rays leave the crystal at an angle $2\theta_B$ with respect to their incident direction. Due to the geometry, all diffracted gamma rays satisfying condition (6.19) for a given energy E and order of reflection n focus onto the same point F, as it is shown in Fig. 6.26.

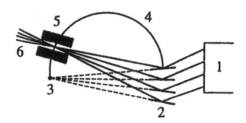

Figure 6.26. Schematic representation of the focussing obtained using a bent-crystal spectrometer in the Cauchois geometry (from [6.29]).

The collection of all these points for all values θ_B lies on a circle with diameter $2R = R_{cyl}$ passing through the centre of the crystal and the point O as indicated in Fig. 6.26. This circle is called the Rowland circle.

Thus, the tuneable γ-ray source is based on the combination of a broad bremsstrahlung source with a curved single-crystal and with a slit system to select the wanted energies.

The Rowland slit (5) performs the energy selection. Behind the Rowland slit there is a quasi-monochromatic beam (energy width is about 1%) ranging from a few tens of keV to a few MeV.

One of the first experiments performed with a tunable gamma-ray source was devoted to studying the feasibility of heavy element tomography [6.30]. When the energy of a monochromatic gamma-ray beam is tuned over the K-edge of an element, the absorption of the beam due to this element changes abruptly. Using this feature of the gamma-ray absorption, energies just above and just below the edge of an element were produced to identify this particular element by means of the absorption characteristics. In the experiments small quantities of uranium were detected and localized first in well-known samples and then in natural stones, without needing to destroy the samples. What is interesting is that not only is the density distribution reported, but a real element identification is performed. The method provides two-dimensional heavy element tomographies.

6.5 Positron Production

The first experiments on positron production were proposed and performed in 1963 by Belovintsev with coworkers [6.31, 6.32] who invented an ingenious scheme allowing the acceleration of positrons in a microtron called the positron microtron. The initial scheme for the positron microtron is shown in Fig. 6.27. At the last orbit the accelerated electron beam is displaced by two short magnetic channels so that they strike the walls of the cavity opposite the emitter. At this point a tungsten electron-to-positron converter 1 mm thick (0.3 rad.lengths) is located behind which a light graphite or aluminium positron attenuator is placed.

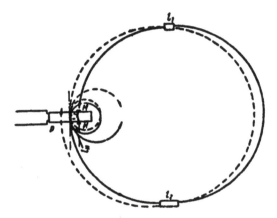

Figure 6.27. Schematic of the positron microtron.

The difference between the lengths of the magnetic channels is equal to the distance of the positron emission point to the axis of the cavity, while the sum of the lengths of the channels is equal to the phase difference which is needed for positron capture into the accelerating mode: $l_2 - l_1 = x_0$, $l_1 + l_2 = \pi, 3\pi$, etc. (all the lengths are expressed in units of $\lambda/2\pi$).

If these conditions are fulfilled, the positron will be captured into an acceleration mode and will move in the opposite direction to the electrons. When the positrons reach the last orbit, the action of the magnetic channels displaces them such that they go through the positron extraction channel. Measurements have shown that the efficiency of electron-to-positron conversion is $(2-5) \times 10^{-6}$ at 6 MeV.

Another design for a positron microtron, also built by Belovintsev at Lebedev Physical Institute, is shown in Fig. 6.28. Essentially, this method involves a double microtron, the right-hand side of which accelerates electrons while the left side accelerates positrons. In this method, it is possible to simultaneously extract an electron beam, a condition which is essential for parallel stacking of positrons and electrons in a storage ring system.

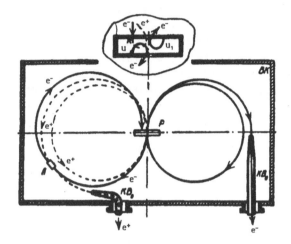

Figure 6.28. The double positron microtron.

The positron beam is just as compact and monoenergetic as the electron beam, and one should keep this in mind when comparing a positron microtron with a linear accelerator in which converted positrons have a much larger spread of energy and transverse momentum. The latter is very essential for an injection of positrons into a synchrotron. For instance, a 460 MeV race-track microtron as an injector of electrons and positrons for LEP (e^+e^- Colliding Beam Machine at CERN) was proposed and designed by Eriksson *et al.* [6.33]. As short (6 ns) pulses of electrons are required, the microsecond pulses from the microtron have to be compressed, and the authors proposed to equip the accumulation ring with an rf system which will work at the sixteenth harmonic.

Actually, the efficiency of electron-to-positron conversion depends on the electron energy. Figure 6.29 represents calculations and experimental data of the total positron yield with electron energy which were performed by different authors and summarized by Kozhemyakin *et al.* [6.34].

The experimentally measured coefficient of electron-to-positron conversion is in rather good agreement with the theoretical calculation. The solid line in Fig. 6.29 corresponds to the efficiency of

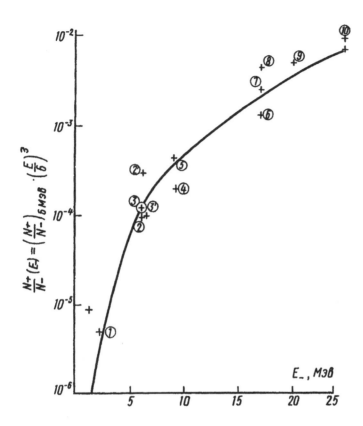

Figure 6.29. Total yield of positrons: 1 – calculation (total yield for all angles from tungsten converter 0.15 rad.length thick, $E_e = 2.0$ MeV) [6.34]; 2, 2' – calculation and experiment for $E_e = 6$ MeV, W-converter, 0.25 rad.length [6.31, 6.32]; 3, 3' – calculation and experiment for $E_e = 6.1$ MeV, 0.15 rad.length [6.34]; 4 – Pt-converter 0.7 rad.length, $E_e = 9.3$ MeV [6.35]; 5 – calculation $E_e = 9$ MeV, W, 0.68 rad.length [6.36]; 6 – $E_e = 15.9$ MeV, Pt, 0.7 rad.length [6.35]; 7 – calculation for $E_e = 17$ MeV, W, 0.68 rad.length [6.36]; 8 – $E_e = 17.9$ MeV, W, 0.2 rad.length [6.37]; 9 – [6.39]; 10 – $E_e = 25$ MeV, W, 0.2 rad.length [6.37].

conversion as electron energy cubed:

$$\frac{N^+}{N_-}(E_-) = \left(\frac{n^+}{N_-}\right)_{6\,\text{MeV}} \cdot \left(\frac{E}{6}\right)^3. \qquad (6.20)$$

Figure 6.30 shows spectra of positrons from tungsten convertor of different thicknesses at electron energy 6.1 MeV. As it is seen, the optimum lies at $0.7 \div 1.0$ mm.

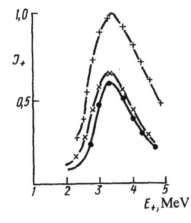

Figure 6.30. Positron spectra with the thickness of a tungsten converter at $E_e = 6.1\,\mathrm{MeV}$ (from [6.34]): $\times$ – 0.2 mm; $+$ – 0.5 mm; $\bullet$ – 2 mm.

Interest in positron beams is not only restricted to nuclear physics. The use of slow positron beams in surface and materials science has expanded at a rapid race during the last decade [6.39–6.41].

This development can be traced partly to an improved understanding of 'moderation', the process by which positrons are reemitted from the surface of materials with negative positron work functions, and partly to the technological developments in the processing and fabrication of the required materials. Current positron beam facilities cover a range of energies from a few eV to several keV and even several MeV.

As for applications, positron beams can be used, for instance, for detailed investigation of the process of positron annihilation, positron channeling radiation studies, which are expected to be sensitive to defects and interstitial impurities, as a depth-sensitive probe for interfaces and multilayered structures in solid materials.

A slow positron source which is based on a microtron was developed by Mills *et al.* [6.43] at the Bell Laboratory in the USA. Energetic shower positrons are produced by the interaction of the relativistic electron beam with the 3.8 mm-thick, sintered W target shown in Fig. 6.31. The positrons are moderated by a W(110) single crystal placed as shown for optimum efficiency. The W crystal is prepared by heating to about 2000 K *in situ*.

The W target is held at ground potential, and the W crystal is biased at +2.75 V. Then the slow positrons from the W crystal are guided to the rf trap and harmonic potential buncher [6.43] which compress the relatively long microtron pulse by three orders of magnitude.

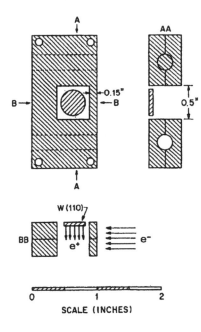

Figure 6.31. Mechanical drawing of the positron production target and slow positron moderator. The electron beam impinges on a target made of sintered tungsten $0.15'' = 3.8$ mm thick. The target is in two halves that are clamped around two $0.25'' = 6.4$ mm-diameter water cooling tubes made of stainless steel. The energetic shower positrons are moderated to a few electron volt energies by a W(110) single crystal that has been prepared by heating up to $\simeq 2000$ K *in situ* (from [6.42]).

The mechanical drawing for the buncher is shown in Fig. 6.32. The buncher is made of 100 rings of stainless steel, with 35.4 mm internal diameter. The rings are connected to two resistor chains.

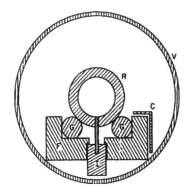
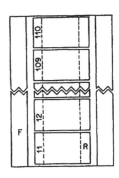

Figure 6.32. Mechanical drawing of the parabolic potential buncher: V – 4-in. external diameter vacuum chamber; R – 304 stainless-steel accelerator ring, 1.00-in. internal diameter, 1.375-in. external diameter; C – mica dielectric capacitor plate; G – 0.500-in. diam quartz insulator rods; F – 304 stainless-steel frame; I – ceramic insulator holding ring in place. There are 100 rings R (from [6.42]).

The positrons are trapped after passing through the pinched magnetic field at the entrance to the magnetic bottle by a transverse rf electric field tuned to the positron cyclotron resonance. Figure 6.33 shows a cross section of the 430 MHz rf cavity for exciting the positron cyclotron resonance. The frequency and the $\approx 1\,V$ amplitude of the rf driving voltage are carefully tuned for optimum trapping.

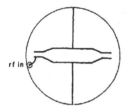

Figure 6.33. Cross section of the 430-MHz rf cavity. The cavity slides into the inside of the 4 in. external diameter vacuum pipe and is 4 in. long (from [6.43]).

Figure 6.34(a) shows the electron beam current versus time as measured by a Faraday cup. At the end of the electron pulse, a high voltage pulse (Fig. 6.34(b)) is applied to the parabolic attenuator accelerator to bunch the positrons. About 100 ns later the positrons arrive at the channel electron multiplier array target with the time distribution shown in Fig. 6.34(c).

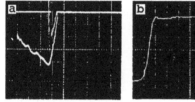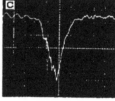

Figure 6.34. *(a)* Electron current versus time; 5 µs and 10 mA per large division. *(b)* Pulse applied to the parabolic buncher; 20 ns and 400 V per division. *(c)* pulse from the plastic scintillator; 10 ns and $\simeq 100\,mV$ per division (from [6.42]).

The best results obtained, using a $1\,mm \times 9\,mm \times 5\,mm$ thick moderator, gave $2 \times 10^5\,e^+$ per bunch initially, the positron pulse width is 7 ns full width at half maximum (fwhm) and the positron current is $\simeq 5 \times 10^6\,s^{-1}$ [6.44]. Since the initial intensity degraded in a few minutes due to heating, in the final version authors used a W single crystal moderator that is held by a water-cooled Mo clamp. After that the yield degrades with a several hour time constant, apparently due to radiation damage, but the yield recovers following an anneal at $\sim 2000\,K$.

Chapter 7

Neutron Sources on the Basis of Electron Accelerators

In the 60's and 70's an intensive construction of nuclear power plants as well as nuclear research reactors took place. Being an intensive source of neutrons, these reactors for the most part served as a basis for the development of applied nuclear physics, in particular for neutron activation analysis. There were also reactors specially designed for some definite application tasks. Thus, in the USSR for Norilsk mining metallurgical enterprise a 100 kW power reactor RG-1M [7.1] was built for the activation analysis of technological products – from the initial raw materials to the output products.

Besides nuclear reactors we now have at our disposal the following nuclear sources:

(i) isotope sources, predominantly ^{252}Cf;

(ii) neutron generators, that are for the most part based on the reaction $p, t \rightarrow {}^3\text{He}, n$;

(iii) proton accelerators with the energy $\sim 1\,\text{GeV}$ — spallation sources;

(iv) electron accelerators for energies 20–$30\,\text{MeV}$.

For electron accelerators neutrons are secondary particles, they originate in the matter as a result of photonuclear reactions (γ, n) and (γ, f). Figure 7.1 shows neutron yields from various targets depending on electron energy which were presented by Gozani [7.2]. An average neutron yield from thin targets which can be considered as a point source of neutrons equals $\sim 10^{10}\,\text{neutr/s}$ at electron energy $\sim 8\,\text{MeV}$ and average current $50\,\mu\text{A}$ (tungsten

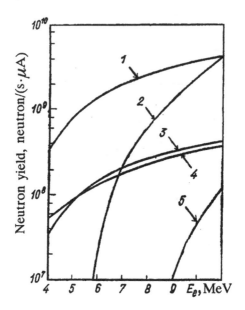

Figure 7.1. Neutron yield from various targets: 1 — D_2O, $t=10$ cm; 2 — ^{238}U, $t=23.8$ g/cm^2 (13 mm); 3 — D_2O, $t=1$ cm; 4 — Be, $t=1$ cm; 5 — W, $t=6$ g/cm^2 (3mm).

(e,γ) converter of optimal thickness, neutron converter ~ 1 cm thick).

A uranium target provides approximately three times higher neutron yield at the same energy. Thick neutron converters of D_2O or Be provide more than an order of magnitude higher yield relative to thin converters (see curve 1 in Fig. 7.1). Certainly, at small electron energies Be and D_2O converters are the best materials, but it is necessary to keep in mind that at higher energies the higher thermal neutron flux density in moderators is provided by compact converters, as it was emphasized in Chapter 6.

The thermal neutron flux density obtained in modern electron accelerators is of course by 2–3 orders of magnitude less than in a standard reactor, but the simplicity of an electron accelerator based neutron source gives rather wide possibilities to experimentalists. Firstly, it is necessary to note that using different converters it is possible to form different fast neutron spectra and to use reactions of the type $(n,n'),(n,p),(n,\alpha)$. If the moderator is made of graphite with a slowing down length ~ 50 cm, a rather extended resonance neutron region can be obtained. There is also another possibility of electron accelerators – a creation of mixed

$(\gamma - n)$-field.

The last but not the least point is that an electron accelerator based neutron source is more universal than a reactor. Besides, taking into account radiation safety and ecology, reactors and accelerators are incomparable installations: an accelerator which can be switched on and off at any time is in this regard without competition.

In this chapter we will consider in detail all scientific and engineering aspects of neutron production at electron accelerators.

7.1 Neutron Slowing Down

Neutrons emitted by neutron sources usually have energies of the order of some keV or MeV. As a result of collisions with nuclei of the matter these neutrons can be elastically or inelastically scattered or captured by nuclei. As the capture cross section of fast neutrons is negligible in comparison with the scattering one, neutrons are moderated and this process occurs mainly an elastic way, as the first excited level is of the order of some MeV for light nuclei, and respectively of the order of some hundred keV for heavy nuclei.

Consider scattering of neutrons in a laboratory system of coordinates (l.c.s.) and in a centre of mass system (c.m.s.) — Fig. 7.2.

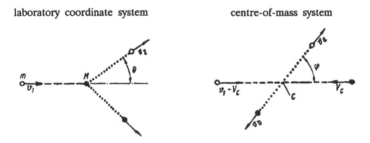

laboratory coordinate system   centre-of-mass system

Figure 7.2. Diagram of neutron scattering by a nucleus in l.c.s. and c.m.s.

A neutron of mass m having a velocity v_1 encounters a nucleus of mass M at rest in l.c.s. The hatched lines in Fig. 7.2. correspond to paths of particles before their impact, dashed lines show their paths after collision. In l.c.s. the neutron is moving after the collision with a velocity v_2 at angle θ in relation to its initial direction. In c.m.s. the center of mass of the system is at rest and a nucleus is moving toward it with a velocity V_c which is a velocity of center of mass in l.c.s. The neutron is moving toward the center of mass with a velocity $v_1 - V_c$. After a collision in the c.m.s. the neutron is moving with a velocity v_a from the center of mass at the angle φ relative to its initial direction; and the nucleus undergoes a recoil in the opposite direction with a velocity v_b. The total momentum in c.m.s. before and after the collision is zero. As it follows from Fig. 7.1

$$v_{\text{nucl}} = V_c = \frac{mv_1}{m+M} \quad , \quad v_n = v_1 - V_c = \frac{Mv_1}{m+M} . \qquad (7.1)$$

The total momentum

$$m\frac{Mv_1}{m+M} - M\frac{mv_1}{m+M} = 0 . \qquad (7.2)$$

It follows from the energy and momentum conservation law that after a collision in c.m.s.

$$mv_a = Mv_b \, , \quad \frac{1}{2}m\left(\frac{Mv_1}{m+M}\right)^2 + \frac{1}{2}M\left(\frac{mv_1}{m+M}\right)^2 = \frac{1}{2}(mv_a^2) + \frac{1}{2}(Mv_b^2) . \qquad (7.3)$$

It immediately follows from Eqs. (7.3) that

$$v_a = \frac{Mv_1}{m+M} = v_1 - V_c \quad , \quad v_b = \frac{mv_1}{m+M} = V_c . \qquad (7.4)$$

Consequently, in the c.m.s. after a collision the velocities differ only by their directions, but not by their values. Reverse transition onto l.c.s. is very easy to do on the basis of the above relations.

Consider the energy loss of the neutron. As it is seen from Fig. 7.2, the maximum possible energy loss ΔE by a neutron with initial energy E_1 occurs in a head-on impact ($\varphi = 180°$). It is simple to obtain that in this case

$$\frac{\Delta E}{E} = \frac{4m/M}{(1+m/M)^2} . \qquad (7.5)$$

Energy loss in an elastic collision leads to the neutron slowing down. As it is seen from Eq. (7.5), the heavier a nucleus the smaller the portion of energy a neutron loses on their impact. Therefore, the best moderator is hydrogen. After some collisions the neutron energy is decreased to such a degree that the neutron turns out to be in thermal equilibrium with nuclei of the moderator and its energy is determined by the temperature of the moderator.

In addition, it follows from formula (7.5) that the maximum possible energy loss (at $\varphi = 180°$) is proportional to the initial energy, and coefficient of proportionality is constant for a given medium. It means that successive neutron energies can be represented in the form

$$E_1 = E_n , \quad E_2 = \alpha E_n , \quad E_3 = \alpha^2 E_n \ldots . \tag{7.6}$$

Thus it is convenient to use a logarithmic energy scale. Let us calculate the mean logarithmic energy loss for a single collision. According to the theorem of cosine it follows from Fig. 7.2 that

$$v_2^2 = v_1^2 \left(\frac{M}{m+M} \right)^2 + v_1^2 \left(\frac{m}{m+M} \right)^2 + 2v_1^2 \left(\frac{M}{m+M} \right)^2 \cos \varphi . \tag{7.7}$$

Therefore, the ratio of neutron energy under successive collisions equals

$$\frac{E_2}{E_1} = \frac{v_2^2}{v_1^2} = \frac{1+r}{2} + \frac{1-r}{2} \cos \varphi , \tag{7.8}$$

where $r = [(M - m)/(M + m)]^2$. The maximum energy loss corresponds to $\varphi = \pi$ and $E_2 = rE_1$.

In c.m.s. scattering is isotropic (at least this is correct for neutrons with energies less than $10\,\mathrm{MeV}$). If σ_s is a scattering cross section, the differential cross section for neutrons scattered isotropically in a solid angle $d\Omega$ equals $\sigma_s(d\Omega/4\pi)$. An element of a solid angle between φ and $\varphi+d\varphi$ equals $2\pi \sin \varphi \, d\varphi = -2\pi d(\cos \varphi)$ and an isotropy of the scattering means that all values of $\cos \varphi$ are of equal probability, and, consequently, all values of the ratio E_2/E_1 from 1 to r are also equiprobable.

Thus, the probability $P\,dE$ for a neutron with initial energy E_1 to change, in a single collision, its energy to a resulting value in the region $(E_2, E_2 + dE)$ is determined by the expression

$$P\,dE = dE/(E_1 - rE_1) . \tag{7.9}$$

Here $E_1 - rE_1$ is the total energy interval to which a neutron can be scattered.

By a definition the mean logarithmic energy loss is

$$\xi = \overline{\ln E_1/E_2} = \overline{\ln E_1 - \ln E_2} . \tag{7.10}$$

According to the definition of a mean value we can write

$$\xi = \int_{rE_1}^{E_2} \ln \left(\frac{E_1}{E_2} \right) P\,dE = \int_{rE_1}^{E_1} \ln \frac{E_1}{E_2} \frac{dE}{E_1 - rE_1} = 1 + \frac{r}{1-r} \ln r . \tag{7.11}$$

For hydrogen $(M/m = 1)$ $\xi = 1$, i.e. on average in hydrogen a neutron after a collision has $1/e$ times less energy. When M/m is large, ξ tends to zero, and a neutron does not practically lose energy. Knowing the value of ξ it is simple to estimate the mean number of collisions for a neutron to slow down to some definite energy. This number equals the total energy on the logarithmic scale divided by ξ. Table 7.1 gives values for some substances of mass number A, parameter ξ and the mean number of collisions needed to moderate a neutron from 2 MeV to thermal energy (0.025 eV).

Table 7.1. Mean logarithmic losses for some substances

Substance	A	ξ	$18.2/\xi$
Hydrogen	1	1	18
Deuterium	2	0.725	25
Helium	4	0.425	43
Lithium	7	0.268	68
Beryllium	9	0.209	87
Carbon	12	0.158	115
Oxygen	16	0.120	152
Uranium	238	0.00838	2172

As distinct from c.m.s. in l.c.s. scattering is not spherically symmetrical, especially for approximating masses m and M. In the latter case neutrons are mainly scattered forward.

Once neutron deceleration is completed, their further motion in a space is described by the diffusion equation. The time variation of the neutron density n at some point of space (x, y, z) is determined by the neutron flux through the boundaries of the volume under consideration, by increasing the number of thermal neutrons and leakage of neutrons due to their capture by atomic nuclei. The most interesting case is a stationary one when the neutron density is independent of time. In this case the diffusion equation has the form

$$\nabla^2 n(x, y, z) - \frac{n}{L^2} + a\frac{q}{L^2} = b\frac{dn}{dt} = 0. \qquad (7.12)$$

Here ∇^2 is Laplacian, q is the velocity of neutron production in 1 cm^3 of the medium, L (diffusion length) and a are constants which are determined by properties of the substance. The term $-n/L^2$ determines neutron leakage due to their capture by nuclei,

aq/L^2 — neutron production, dn/dt — neutron density change in time, and $\nabla^2 n$ is a balance of neutron fluxes.

As an example we consider neutron diffusion from a point neutron source. In the case considered the value n depends only on r, that is the radial distance from the source, $q = 0$ (there is not neutron production in the media), and thus the Laplacian has the form

$$\nabla^2 n = \frac{d^2 n}{dr^2} + \frac{2}{r}\frac{dn}{dr}. \qquad (7.13)$$

We insert a new variable $F = nr$. Then the diffusion equation is in the form

$$\frac{d^2 F}{dr^2} - \frac{F}{L^2} = 0, \qquad (7.14)$$

which has the solution

$$F = Ce^{-r/L} \quad \text{or} \quad n = \frac{C}{r}e^{-r/L}. \qquad (7.15)$$

The factor C in the solution (7.15) is determined by the intensity of the point source. Note, that it follows from the solution (7.15) that $\ln[rn(r)]$ is a linear function of r:

$$\ln[rn(r)] = -r/L. \qquad (7.16)$$

By means of this expression one can determine the diffusion length.

7.2 Spatial Distribution of Neutrons from a Point Source

In the case of a point source of fission neutrons placed in infinite homogeneous media, the theory gives the following formula for spatial distribution of thermal neutrons

$$\Phi_{th}(r) = \frac{QL^2}{4\pi Dr}\left(\frac{e^{-r/L}}{L^2 - \tau_{th}} - \frac{e^{-r/\sqrt{\tau_{th}}}}{L^2 - \tau_{th}}\right), \qquad (7.17)$$

where Φ_{th} is the thermal neutron flux density, n/(cm^2·s), Q is the neutron flux from the source, n/s, D is the diffusion coefficient, cm^{-1}, and τ_{th} is the neutron age for slowing down from the initial up to thermal energy, cm^2 (in this particular case, when the final energy is the thermal one, neutron age equals slowing down length squared).

The neutron parameters needed for the calculation, are listed in Table 7.2.

Table 7.2. Neutron parameters of different moderators with a fission neutron source

Moderator	Neutron age τ_{th}, cm^2	Diffusion length L, cm	Diffusion coefficient D, cm^{-1}
Water	28	2.8	0.16
Beryllium	90	21.2	0.50
Graphite	368	52.5	0.86

Figure 7.3 displays radial distribution of thermal neutrons in water, beryllium, and graphite, calculated in accordance with formula (7.17). The open circles in Fig. 7.3 are experimental results confirming the validity of the calculations.

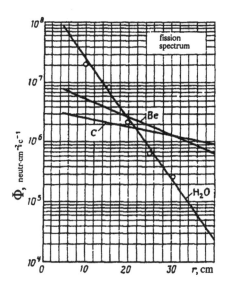

Figure 7.3. Spatial distribution of thermal neutrons from a fission fast neutron source in water, beryllium and graphite; points — experimental results, solid lines — calculations (from [7.2]).

7.3 Neutron Converters and Moderators

In the simplest version the neutron moderator is water or a polyethylene cube in whose center a fast neutron source is placed. To produce neutrons in electron accelerators, a high-energy electron beam is introduced into the moderator set-up by a vacuum electron guide in whose end a water cooled bremsstrahlung target (tungsten or tantalum 4-5 mm thick) and a gamma-to-neutron converter (for example lead 5 cm in length) are located. Below we will consider experimental results [7.3] which are useful for correctly designing a neutron source in electron accelerators.

Consider first the properties of a fast neutron converter placed behind a bremsstrahlung target. Neutron and gamma fluxes were measured by means of threshold detectors. To detect fast neutrons the following reactions were used: $^{27}Al(n,p)^{27}Mg$ $(T_{1/2} = 9.5\,min,$ $E_\gamma = 0.84$ and $1.015\,MeV$, $E_{thr} = 1.89\,MeV$), $^{27}Al(n,\alpha)^{24}Na$ $(T_{1/2} = 15\,h$, $E_\gamma = 1.37$ and $2.76\,MeV$, $E_{thr} = 3.26\,MeV$).

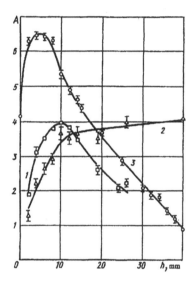

Figure 7.4. Dependence of fast neutron flux on the lead converter thickness measured by aluminium detectors: 1 – detectors are placed at the angle $\theta = 0°$ relative to electron beam axis, 2 – $\theta = 180°$, 3 – detectors are along the axis of the converter.

It is obvious that the transverse size of a neutron converter is determined by the electron beam size and by divergence of γ-radiation, and its longitudinal size has to be chosen in such a way as to provide the maximum use of γ-radiation. We consider that the neutron converter is optimal if it provides the maximum neutron flux. The dependence of fast neutron flux on the lead converter thickness is shown in Fig. 7.4 for 0° and 180° relative to the electron beam axis. It is seen that the optimal thickness of a lead neutron converter equals ∼ 10 mm (∼ 1.7 rad.length). A converter of such thickness makes it possible to use gamma-quanta and neutrons simultaneously in an activation experiment.

Curve 3 in Fig. 7.4 shows the distribution of fast neutron flux along the element of the 35 mm thick cylinder converter (6 rad.length). Lead converter of such a thickness provides practically the maximum value of integral neutron flux. The maximum of fast neutron flux is observed at a distance of 5–6 mm. The proximity of the neutron formation zone to the forward edge of the converter is due to the high value of γ-quanta absorption in lead.

To measure the characteristics of thermal neutron field a $600 \times 600 \times 600$ mm cube with distilled water was used. Lead converter 55 mm in diameter and 35 mm thick is placed in the center of the cube. The electron beam is transported to the bremsstrahlung target (2 mm tungsten in the forward side of the converter) through a vacuum tube 60 mm in diameter. Figure 7.5 shows the behaviour of fast and thermal neutron flux in water with the distance from the converter center. As a thermal neutron detector the organic sodium salt $C_6H_4OHCOONa$, sealed in a polyethylene film, was used. In a result of a (γ, n)-reaction the radionuclide ^{24}Na is formed whose γ-activity is measured in the energy interval $E_\gamma = 1.37\text{–}2.76$ MeV.

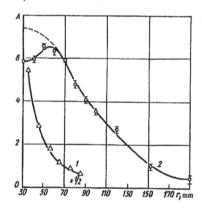

Figure 7.5. The distribution of fast (Δ) and thermal (o) neutrons in the water moderator; dashed line corresponds to the thermal neutron distribution in a water moderator for a point neutron source.

At a distance from the converter $r > l$, where l is the characteristic size of a converter, the latter can be considered as a point source, and the thermal neutron flux behaviour in the moderator with a distance from the source of $r > l$ should be analogous to the well investigated case of a point source with a neutron fission spectrum. Actually, the difference between point source behaviour and experimental measurements is observed only at small distances. Besides, during the experiments the maximum thermal neutron flux is observed at approximately 25 mm from the converter. The appearance of the maximum is connected, of course, with the finite size of the source, and, on the other hand, with the leakage of thermal neutrons through the electron beamline.

In the angular distribution of fast, as well as thermal, neutrons measured at the distance 90 mm from the neutron converter, a small anisotropy is observed $(P(90°)/P(0°)=1.15)$. This indicates that besides evaporated (isotropically) neutrons there are a small number of neutrons reflecting the dipole character of γ-quanta absorption by nuclei, i.e., neutrons emitted predominantly at a right angle relative to the incoming beam axis.

To determine the optimal size of the moderator, the thermal neutron flux behaviour with the thickness of the paraffin moderator in a 2π-geometry was measured. Thermal neutrons were detected by metallic aluminium and manganese detectors, and the activity of the nuclides ^{28}Al $(E_\gamma =1.78\,\text{Mev})$ and ^{56}Mn $(E_\gamma=0.84, 1,81$ and $2.1\,\text{MeV})$ were measured.

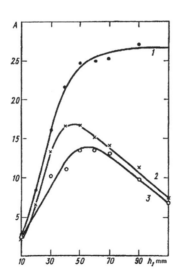

Figure 7.6. Thermal neutron flux behaviour with a thickness of paraffin moderator measured at its front (1) and back (2,3) side by aluminium (o) and manganese (•, ×) detectors.

Figure 7.6 shows the aluminium and manganese activity measurements, placed behind the paraffin moderator, depending on its thickness. It follows from the results that the maximum thermal neutron flux is obtained at the paraffin moderator 45-50 mm thick. A small difference in the moderator thickness, observed with aluminium and manganese detectors, is due to the different contribution of epithermal neutrons into the total activity of the detectors used.

The dependence of the activity of manganese detectors, placed at the front of the moderator, is also shown in Fig.7.6 with its thickness. It follows from the figure that the paraffin moderator of 110 mm thickness (two moderation lengths) practically provides the maximum thermal neutron flux density.

The angular distribution of the bremsstrahlung from the target with the neutron converter is presented in Fig.7.7. Measurements were carried out by means of copper and carbon (polyethylene) detectors that register γ-quanta with energies $E_\gamma > 10.8$ and $E_\gamma > 18.6$ MeV respectively. It is seen that the presence of a moderator leads to a considerable spread of the bremsstrahlung, and a fraction of γ-quanta with $E_\gamma > 10.8$ and $E_\gamma > 18.6$ MeV at the angle $\theta = 90°$ is 28 and 2.8% in relation to the intensity of bremsstrahlung in the forward direction ($0°$). The lower the energy of γ-quanta, the more symmetric their distribution becomes.

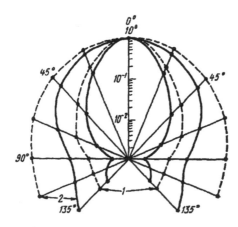

Figure 7.7. Angular distribution of the high-energy part of the bremsstrahlung from the lead target: 1 – lead target 55 mm in diameter and 35 mm thick; 2 – the same target with water moderator. Solid line corresponds to copper detectors, dashed line to carbon detectors.

As a rule, for neutron activation analysis small pockets of several centimeters in size are made in a moderator. However, the possibility of making holes of different sizes in a moderator could broaden significantly the range of problems to be solved. This applies particularly to analysis of extended objects and liquids, which is rather complicated, or even impossible to perform in reactors. For instance, the following ranges of problems can be solved with a large irradiation pocket:

1) Determination of the content of inorganic impurities in oil and processed products. Resolving this problem allows us to evaluate the possibility of rare metal production from oils as well as to estimate the level of atmospheric pollution due to the presence of toxic elements in a fuel. It should be emphasized that the major disadvantage of chemical and physico-chemical methods in solving such problems lies in the need to ash samples with the inevitable contamination by microimpurities. Moreover, ashing causes the loss of volatile elements such as B, Cl, V, As, Sb, and I etc. Besides, oil is a combustible liquid and activation analysis of such an object in reactors is always a difficult task due to safety problems.

2) Nondestructive determination of noble metals in electronic scrap, such as boards and their electronic constituents. The main interest is to analyse the gold, platinum, palladium and silver content to assess the feasibility of their extraction.

A hole in the moderator inevitably leads to distortion of the neutron distribution that exists in an infinite homogeneous moderator. Moreover, the distance required for themalization of fast neutrons in the hydrogen containing moderator is only $\sim 5\,cm$, and therefore the thermal neutron flux drops heavily when moving away from the source, i.e. the neutron field would be very non-uniform in the cavity.

However, the formation of the thermal spectrum in a large cavity is a result of multiple reflections from the moderator walls and permanent back flux from the moderator to the cavity. As a result, even in a large cavity a rather homogeneous neutron field is created with an almost thermal spectrum, i.e. a thermalized 'neutron gas' is formed in the cavity. This was confirmed by direct measurements in the 30 Mev microtron in IPP [7.4].

The water-cooled $Ta(3\,mm)+Pb(60\,mm)$ target 50 mm in diameter was used as a bremsstrahlung target and photoneutron source. The target was placed in the centre of a polyethylene cube $60 \times 60 \times 60\,cm^3$ with a cylindrical hole of 15 cm diameter in its upper side, as it is shown in Fig. 7.8. The cylindrical hole is closed by a movable polyethylene cover that allows the free space in the moderator to be varied up to $2500\,cm^3$. The radial thermal neutron flux distributions in different parts of the cavity, also shown in Fig. 7.8, indicate a practically uniform neutron field, and there is only a little reduction of flux near the converter.

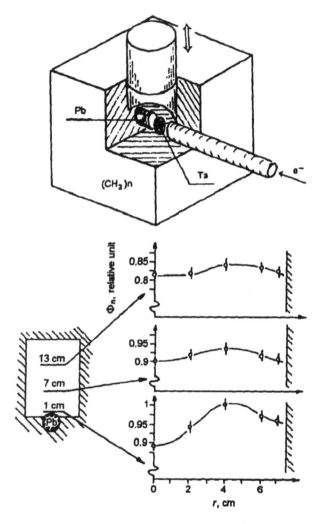

Figure 7.8. Scheme of a moderator with a large-volume hole and radial distribution of thermal neutrons at different heights (1, 7 and 13 cm) in the cylindrical cavity of 20 cm height and 15 cm diameter.

7.4 Resonance Neutrons

Neutron moderators made of light elements provide the maximum available thermal neutron flux. On the other hand, to obtain

a sufficiently wide region of resonance neutrons in a moderator
it is desirable to use a moderator made with elements heavier
than hydrogen. The most appropriate material for this purpose is
graphite. Such a device was developed for the Dubna microtron
[7.5, 7.6], and below we consider the main characteristics of this
moderator which is successfully used for activation analysis (see
Part III).

One of the major parameters of the irradiation device is the
density of the thermal neutron flux that is primarily determined
by the energy and intensity of the neutron source and by the
geometrical sizes and purity of the moderator, and this density
can be calculated analytically. The sizes of the graphite cube,
designed for activation with resonance neutrons, are chosen so
the neutron flux in the operating region of irradiation is slightly
differed from the neutron density in a moderator of infinite size.
The distribution of neutron flux $\phi_{r\infty}$ can be calculated by the
formula obtained from the solution of the neutron age equation
and given in [7.7]

$$\phi_{r\infty} = \frac{Q}{\xi \Sigma_s (4\pi\tau)^{3/2}} \, e^{-r^2/4\tau}, \qquad (7.18)$$

where Q is the number of neutrons per second emitted by the
source, r is a distance from the source, ξ is the logarithmically-
averaged energy loss per one act of scattering, τ is the neutron
age under its moderating from the initial photoneutron energy up
to resonance energy, and Σ_s is the macroscopic scattering cross
section. The neutron age can be calculated by the expression

$$\tau = \tau_{\text{th}} \frac{\lambda_s^2}{3\xi(1 - 2/3\,A)} \int\limits_{E_{kT}}^{E} \frac{dE'}{E'}, \qquad (7.19)$$

where $\tau_{\text{th}} = 350\,\text{cm}^2$ is the effective neutron age under neutron
of fission spectrum slowing down to thermal energy, λ_s is the
neutron free path in relation to scattering, A is the atomic weight
of moderator nuclei, and $E_{kT} \simeq 3.5\,kT$ is the boundary energy of
the thermal region accepted.

Let the operating region be a distance r_0, where the density
of resonance neutron flux is halved in relation to its maximum
value. In such a case $r_0 = 30\,\text{cm}$ for the neutron energy $4.9\,\text{eV}$.
For the moderator of cubic shape with a side R in the center
of which the fast neutron source is situated, the expression for

neutron flux has the form

$$\phi_r = \frac{8Q}{\xi\Sigma_s R_1^3} \sum_{l,m,n=1}^{\infty} \exp[-(\pi/R_1)^2(l^2+m^2+n^2)\tau]\times$$

$$\sin^2\frac{m\pi}{2}\sin^2\frac{n\pi}{2}\sin\frac{l\pi}{2}\sin\frac{l\pi x}{2R_1}, \qquad (7.20)$$

where $R_1 = R + 2d$, d=1.9 cm is the extrapolated boundary of the media, i.e., the distance from the moderator surface where the extrapolated value of the function ϕ_r is zero. The results of the calculations and the experimental data are shown in Fig. 7.9.

Similar calculations have been carried out for the thermal neutron flux [7.9]. The infinite moderator and the cube shaped moderators with sides of 60, 80, 100 and 120 cm were examined and the results are shown in Fig. 7.9. An additional assumption in these calculations was that the macroscopic cross section in the neutron thermalization process was equal to zero.

The experiments, carried out in order to measure the thermal and epithermal neutron flux distributions, fully confirmed the dependences shown in Fig. 7.9.

Comparing the experimental and calculated data we can come to the following conclusions:

(i) The optimum size of the graphite moderator for resonance neutrons is a cube with sides 100 cm, and a further increase of its dimensions does not result in an increase of the neutron flux density in the working area.

(ii) Theoretical formulae (7.18) and (7.20) describe well the experimental data, except for the cube with sides of 60 cm.

(iii) Increasing the cube side length from 60 to 120 cm leads to increasing thermal neutron flux density and it reaches approximately one half of its value for a moderator of infinite size; only a 2 m-side cube has a density close to the maximum value.

The results given are obtained for a homogeneous moderator, but in practice it is necessary to have irradiation channels that definitely influence the neutron flux.

The neutron flux density can be increased in the central part of the cube by surrounding the converter with light-element material (initial moderator). It was shown experimentally that the neutron flux density increases by 1.5 times by using an initial polyethylene moderator 5–10 cm thick, or by using a 4 mm 'water jacket', which cools the converter.

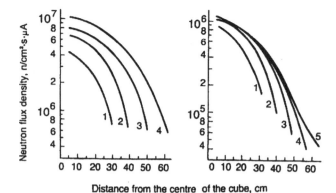

Distance from the centre of the cube, cm

Figure 7.9. *Left*: Spatial distribution of epithermal neutron flux density for various moderator dimensions: 1 — $R = 60$ cm; 2 — $R = 80$ cm; 3 — $R = 100$ cm; 4 — $R = 120$ cm; 5 — infinite moderator. *Right*: Spatial distribution of thermal neutron flux density for various moderator dimensions: 1 — $R = 60$ cm; 2 — $R = 80$ cm; 3 — $R = 100$ cm; 4 — $R = 120$ cm.

Better results were obtained by using (close to the converter) a second moderator made of aa lighter material (a so-called initial moderator). It was shown experimentally that the best results are obtained with a cuboid initial beryllium moderator with sides $12 \times 50 \times 50$ cm^3.

In addition to moderation, beryllium increases the efficiency of $\gamma \rightarrow n$ conversion due to the low energy threshold of (γ, n)-reaction. The employment of an initial beryllium moderator increases neutron flux density in the working area approximately at a twofold rate. Figure 7.10 shows an influence of various initial moderators on the resonance neutron flux density in a graphite cube with sides 120 cm.

To obtain the greatest neutron flux density in the operating device, natural uranium in the shape of a cylinder 35 mm in diameter and 30 mm in length was used as a neutron converter (the uranium was hermetically packed in a water cooled aluminium container). Substitution of the uranium converter for a beryllium one results in the flux density increasing, and correspondingly detection limit by 2.5 times.

Figure 7.11 presents the schematical view of the graphite moderator with a uranium-beryllium converter in its center. The electron beam bombards the uranium converter (cylinder 30 mm in diameter and 30 mm in length) surrounded by beryllium $500 \times 500 \times 120$ mm in size. The thermal and resonance neutron distri-

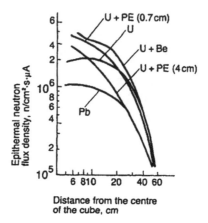

Distance from the centre
of the cube, cm

Figure 7.10. Influence of various kinds and thickness of initial moderators (beryllium, polyethylene (PE), uranium, lead) on the epithermal neutron flux density in the graphite cube with side length of 120 cm.

butions in this device are also shown in Fig. 7.11. The epithermal (resonance) neutron flux density equals $2 \cdot 10^7 \, \text{n/cm}^2 \cdot \text{s}$ at electron energy 16 MeV and average current 25 μA.

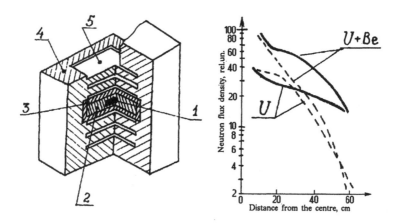

Figure 7.11. *Left:* The graphite cube for resonance and thermal neutron activation analysis: 1 — uranium target, 2 — beryllium, 3 — lead, 4 — graphite, 5 — channels for irradiation of samples. *Right:* Neutron distribution in the graphite moderator with U- and (U+Be)-converters: solid line – thermal neutrons (0.025 eV), dashed line – resonance neutrons (4.9 eV).

Part III. APPLICATIONS
Chapter 8

Nuclear Reactions Induced by Gamma-Quanta

We consider in this chapter only nuclear reactions induced by γ-quanta, which are called *photoneutron reactions*. To describe the process of nuclear reconstruction under the action of an incoming particle, we have to know the properties of nuclei in the ground and excited states, as well as the conservation laws that govern nuclear reactions.

8.1 Classification of Photons and Gamma Radiation

As a photon is a zero-mass particle (it is essential also that its spin equals 1), it is impossible to use for it the concept of angular momentum, and instead the idea of multipolarity is put forward. The electromagnetic field multipole is a state of a free propagating field that is characterized by total momentum and parity. The state with momentum L and parity $(-1)^L$ is called *electric 2^L-pole* and designated by letter E, and the state with momentum L and parity $(-1)^{L+1}$ is *magnetic 2^L-pole* (designated by M). The low states have special names: *dipole* $(L = 1)$, *quadrupole* $(L = 2)$, and *octupole* $(L = 3)$. As a rule, there is always γ-quantum emission of the lowest multipolarity. This rule is connected with the fact that increasing multipolarity by one leads to decreasing the emission probability, i.e. to decreasing the γ-transition probability by $(\lambda/R)^2$ times, when R is nuclear radius and $\lambda = \lambda/2\pi$ is a wave length of radiation. In addition, being equal multipolarity the probability of magnetic radiation is also (λ/R) times less than electric radiation. Thus, displacing transitions in decreasing order of their radiation probabilities, we obtain a set: $E1$, $E2$ and $M1$, $E3$ and $M2$, and so on.

As it is known from electrodynamics, the power of dipole radiation equals

$$W = \frac{1}{3c^3}\omega^4 d^2 ,$$ (8.1)

where d is the dipole moment. It means that one γ-quantum with the energy $E = \hbar\omega$ is emitted in such a time τ that

$$\frac{1}{\tau} = \frac{W}{\hbar\omega} = \frac{d^2}{3\hbar c^3}\omega^3 .$$ (8.2)

So, the probability of dipole transition is proportional to its energy cubed. In other words, as in accordance with the uncertainty principle, the level width and life-time are connected by the relation $\Delta E \cdot \tau \simeq \hbar$, then the radiative (dipole) level width is

$$\Gamma_\gamma \propto E^3 .$$ (8.3)

For most nuclei the dipole transition time is $10^{-13} - 10^{-17}$ s.

Photon emitted in either electric or magnetic 2^L pole radiation carries away an amount of angular momentum equal to the vector $\vec{L}$ of absolute magnitude $\sqrt{l(l+1)}\hbar$, just in the case of a particle. The rank L defines the multipolarity of the photon as the number of units of angular momentum removed by the quantum emission, since angular momentum is a conserved quality in electromagnetic interactions. This implies for the quantum numbers

$$|J_i - J_f| \geq L \geq J_i + J_f ,$$ (8.4)

where J_i, J_f are the initial and final angular momentum of the nucleus. Equation (2.48) means that L can have any integral value from $|J_i - J_f|$ to $(J_i + J_f)$. This constitutes the momentum selection rule which curbs the permitted range of multipolarities L.

The life-time of a majority of excited nuclei that are decayed by γ-emission is too short in comparison with the lifetimes of charged particle decays, and usually lies in the range $10^{-7} - 10^{-11}$ s. In some rare cases for high order forbidden transitions with small amounts of energy released, one can observe gamma radioactive nuclei with a macroscopic lifetime of an order of some hours and even of some years. Such long-living excited nuclear states are called *isomers*. The isomer level would have a spin that is very different from low-lying states. As a rule, such isomer states belong to first excitation nuclear levels and are usually the nuclei with almost magic nucleon numbers 50, 82, 126 (see Fig. 8.1).

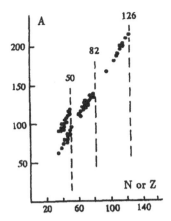

Figure 8.1. Distribution of isomers with $T_{1/2} > 1\,\mathrm{s}$ among nuclides with odd A.

In these regions the shell levels are close to each other in energy but vary greatly in spin values.

8.2 Features of Photonuclear Reactions

Nuclear reactions under the action of γ-quanta with energies up to $30\,\mathrm{MeV}$ have the following peculiarities. Firstly, as electromagnetic interaction weaker than nuclear interaction by three orders of magnitude, these processes proceed with less intensity than other nuclear reactions. Secondary, an electromagnetic process in a nucleus follows both electromagnetic and nuclear interactions. Thirdly, at the same energy a reduced photon wavelength λ_γ is much larger a nucleon wavelength λ_N of the same energy

$$\lambda_\gamma = \frac{2 \cdot 10^{-11}}{E_\gamma} \ (\mathrm{cm}), \quad \lambda_N = \frac{4.5 \cdot 10^{-13}}{\sqrt{E_N}} \ (\mathrm{cm}). \qquad (8.5)$$

It follows from (8.5) that while λ_N reaches nuclear size at the energy $\simeq 1\,\mathrm{MeV}$, the corresponding photon energy equals $\simeq 30\,\mathrm{MeV}$. That is why photonuclear reactions predominantly take place under dipole absorption of γ-quanta by nuclei.

Figure 8.2 represents schematically the energy behaviour of photoabsorption cross section σ_γ.

In the energy region I, below the threshold of (γ, n)-reaction only elastic and inelastic scattering of γ-quanta are possible; the cross section exhibits some maxima corresponding to the transitions between discrete levels of a target nucleus. In region II

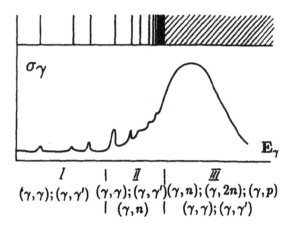

Figure 8.2. Schematic energy behaviour of γ-quanta absorption cross section by idealized nucleus: in the top the level spectrum of the nucleus is shown.

a particle escape is energetically allowed but the level spectrum remains discrete and this is reflected in the fine structure of the cross section. Region III corresponds to overlapping levels of the compound nucleus. In this region photonuclear cross sections display a wide maximum (giant resonance) in the γ-quantum energy region $10-20\,\text{MeV}$.

Giant resonance corresponds to a collective degree of freedom, namely vibrations of the whole neutron mass relative to all protons, that are called dipole nuclear vibrations. The A-dependence of the giant resonance frequency is easily estimated. For every oscillator the resonance frequency ω_0 is determined by a stiffness k and vibrating mass m $(\omega_0 = \sqrt{k/m})$. In the considered mechanism of dipole vibrations (the so-called *Goldhaber-Teller model*) the role of the elastic restoring force is played by the interaction of 'shifted' nucleons with the nucleus (see Fig. 8.3). The number of such nucleons is proportional to the value of the nuclear surface, namely R^2, and the mass of vibrating nucleons is proportional to R^3. Hence, we obtain the following relation for the frequency, and consequently the energy of giant resonance

$$\omega_d = \sqrt{k/m} \propto \sqrt{R^2/R^3} \propto R^{-1/2} \propto A^{-1/6}. \qquad (8.6)$$

In the *Steinwedel-Jensen model* of giant resonance, the proton and neutron fluids are two compressible interpenetrating fluids moving within the rigid surface of the initial nucleus. Then the incident photon field sets up two different evolutions for the proton density ρ_p and the neutron density ρ_n inside the nucleus, whereas the total mass distribution of the nucleons is not perturbed by the vibrations of the proton and the neutron fluids. The heart of the theory is that there is a restoring force, proportional to the symmetry term E_S in the Weizsäcker mass formulae, which tends to restore the normal value of ρ_p and ρ_n. This means that in resonance vibrations the length of the circumference of the nucleus or its diameter has to be equal to the integer number of wavelengths. Therefore the energy of the fundamental dipole mode is predicted by the model as

$$\omega_d \propto 1/R \propto A^{-1/3}. \qquad (8.7)$$

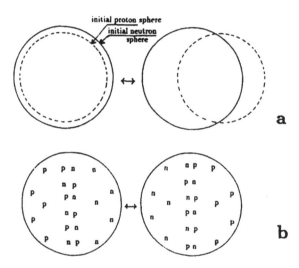

Figure 8.3. The schematic representation of nuclear dipole vibrations: a – in the G.T. model, and b – in the S.J. model; neutron and proton distributions are shown in two different phases of vibration.

These two hydrodynamical models of nuclear dipole vibrations predict $A^{-1/3}$ and $A^{-1/6}$ dependence of giant dipole resonance. Experimental data of the maxima of giant resonance for different nuclei are represented in Fig. 8.4 in the coordinates $EA^{1/3}(EA^{1/6})$ and A. One can see from this presentation that experimental data, describing well the general trend, fit neither, on the other hand, the $A^{-1/3}$ ($S.J.model$) not the $A^{-1/6}$ law ($G.T.\ model$) very well.

It is now considered preferable to approximate the position of the resonance energy of dipole giant resonance as

$$E_d = 32a^{-1/3} + 21A^{-1/6}. \qquad (8.8)$$

The giant resonance in the photoabsorption cross section is usually approximated by a Lorentzian (with two Lorentzians for deformed nuclei)

$$\sigma_\gamma = \sigma_0 \frac{(\Gamma E)^2}{(E^2 - E_0^2) + (\Gamma E)^2},\qquad(8.9)$$

where E_0 is the resonance energy, Γ — its width ($\sim 2.5 \div 3.5\,\mathrm{MeV}$), σ_0 — the cross section value in the maximum.

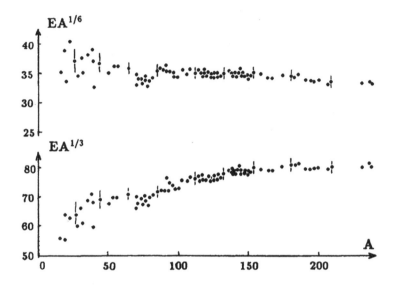

Figure 8.4. Experimental results of the average energy E_d of the giant dipole resonance versus the mass number A (from [8.1]).

Deformed nuclei display splitting of giant resonance (see Fig. 8.5) that corresponds to excitations along and transverse to the axis of nuclear symmetry. The validity of this explanation of resonance splitting was confirmed by measuring the photoneutron cross section for oriented nuclei.

The larger the nuclear mass, the less the neutron binding energy B_n, giant resonance frequency and the larger the absolute value of the cross section σ_0. That is why at the same energy of incoming γ-quantum yields of (γ, n)-reactions differ for different nuclei by many orders of magnitude (see Fig. 8.6).

The behaviour of giant resonance parameters is a consequence of the dipole character of γ-quantum interaction with nuclei. The theoretical consideration of dipole transitions in nuclei shows that

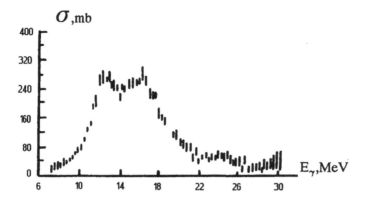

Figure 8.5. Total cross section of (γ, n)-reaction for ^{160}Gd.

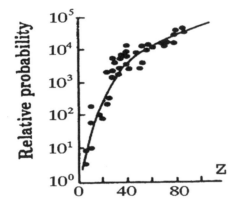

Figure 8.6. Relative yield of (γ, n)-reaction for different nuclei measured with bremsstrahlung with $E_{max} = 20$ MeV.

an integral cross section of nuclear photoeffect has to satisfy the Levinger-Bethe relation

$$\sigma_{int} = \int\limits_{E_{thr}}^{\infty} \sigma(E)\,dE = 0.06\,NZ/A, \qquad (8.10)$$

i.e., it has to be proportional to the nuclear charge Z. A comparison of the observed total photoabsorption cross section (up to 30 MeV) with theoretical prediction shows that the dipole absorption actually brings about the main contribution to the region of giant resonance (see Fig. 8.7).

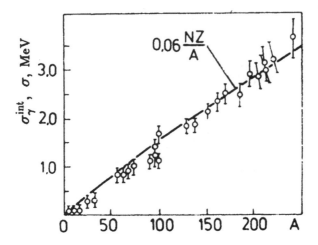

Figure 8.7. Integral cross section of photon absorption in dependence on Z of nuclei and its limiting value according to Levinger–Bethe relation.

8.2.1 Decay Channels

Neutron escape is in most cases a predominant decay channel of a nucleus excited in the region of giant resonance [8.2]. A statistical model describes well the characteristics of the main neutrons: the neutron energy is about nuclear temperature, and their energy spectrum follows the evaporation spectrum

$$n(\varepsilon)\,d\varepsilon \propto \sigma(\varepsilon)\,\varepsilon\,\rho(E^* - B_n - \varepsilon)d\varepsilon, \qquad (8.11)$$

where ε is neutron energy, ρ is the state density of a final nucleus, and σ is the cross section of the inverse process. In the high-energy part of the neutron spectrum the minor ($\sim 5\%$) contribution from neutrons due to direct interaction is also observed. Using the expression for level density, formula (8.11) can be rewritten as

$$n(\varepsilon) \propto \varepsilon\,\sigma(\varepsilon)\exp\left(-\varepsilon/\kappa T\right). \qquad (8.12)$$

The concurrence between neutron and proton escape from an excited nucleus depends on the height of the Coulomb barrier for a proton, the difference of their binding energies, and the level density of the final nuclei. For most medium nuclei ($10 < A < 30$) the proton binding energy is found to be less than a neutron one, as a neutron emission leads to positron-active nucleus, whereas its stable isobar is formed after a proton ejection. Hence, the relative probability of proton emission is rather high for medium mass nuclei although it fluctuates from nucleus to nucleus.

Figure 8.8 displays data on the relative probability of proton emission under bremsstrahlung with maximum energy 23 MeV. Vertical lines correspond to experimental values and the solid line illustrates the statistical model calculation. One can see a good description of experiments up to $A = 45$, but then for heavy nuclei a deviation is observed that is explained by the increasing probability for direct processes (dashed line in the figure).

The statistical model describes well the energy spectrum of photoprotons from medium mass nuclei — see Fig. 8.9. In the proton energy spectrum we also see a small tail in the region of high energies that is ascribed to direct photoeffect. The maximum in the proton energy spectrum corresponds to the energy that is slightly lower than the height of the Coulomb barrier $(Z/A^{1/3})$ and for medium nuclei is 4–5 MeV.

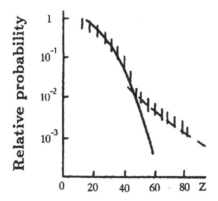

Figure 8.8. Relative probability of proton emission from different nuclei at $E_{\text{max}} = 23$ MeV.

If the energy of γ-quantum is less than the nucleon binding energy, a scattering of photons is the only possible nuclear process [8.3]. In the case of discrete nuclear levels the photoabsorption cross section on one level is described by the following formulae

$$\sigma_a = 2\pi\lambda^2 \frac{2J+1}{2J_0+1} \frac{\Gamma_0}{\Gamma} \frac{(E\Gamma)^2}{(E_0^2 - E^2)^2 + (E\Gamma)^2}. \qquad (8.13)$$

Here J and J_0 are spins of excited and ground states, Γ_0 – radiative width of a transition to the ground state, Γ – the total width of the level.

On the other hand, the integral absorption cross section for a given level depends only on the transition width to the ground

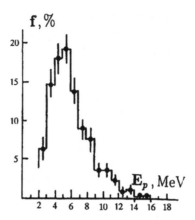

Figure 8.9. Photoproton spectrum from ^{58}Ni at $E_{max}=24$ MeV.

state and on the energy of the incoming proton

$$\int \sigma_a(E)\, dE = (\pi\lambda)^2 \Gamma_0 \frac{2J+1}{2J_0+1}. \qquad (8.14)$$

Therefore, the integral cross section for elastic scattering (in particular, this value is measured with bremsstrahlung) equals

$$\sigma_s^{el} = \int \sigma_s(E)\, dE = \frac{\Gamma_0}{\Gamma} \int \sigma_a(E)\, dE = (\pi\lambda)^2 \frac{2J+1}{2J_0+1} \frac{\Gamma_0^2}{\Gamma}. \qquad (8.15)$$

Most nuclei have high density of states and inelastic scattering of γ-quanta takes a dominant role when deexcitation occurs through a cascade of γ-quanta. In some cases inelastic scattering of photons leads to an isomeric state, the spin of which is very different from the spin of the ground state.

Analogously to elastic scattering the integral cross section for population of isomeric state with spin J_m equals

$$\sigma_s^{isom} = (\pi\lambda)^2 \frac{2J_m+1}{2J_0+1} \frac{\Gamma_0^2}{\Gamma}. \qquad (8.16)$$

So, the ratio of the number of transitions through isomeric state to the total number of γ-transitions, i.e., the probability of isomer state population, equals

$$\frac{\sigma_s^{isom}}{\sigma_s} = \frac{2J_m+1}{(2J_m+1)+(2J_0+1)}. \qquad (8.17)$$

A typical example of isomer state formation cross section (nuclide ^{89}Y) is shown in Fig. 8.10. The corresponding integral yield with bremsstrahlung is also given in the figure. The energy of the isomeric state ^{89m}Y is equal to 909 keV, and its half-life is 16.1 s.

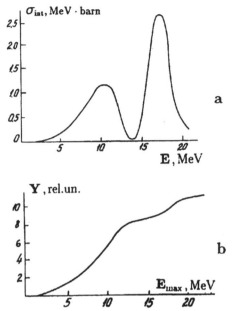

Figure 8.10. a — Cross section of ^{89}Y formation in the reaction (γ, γ'). b — Integral yield of the reaction (γ, γ') at nuclide ^{89}Y.

At low energies of γ-quanta an increase of the cross section is due to increasing radiative width Γ_0. However, above the threshold of (γ, n)-reaction the cross section falls off sharply due to competition with neutron emission (a sharp rise in the total width of excited state). This tendency is also clearly seen from Fig. 8.11, where the excitation function of hafnium-18 and osmium-19 are shown. Maximuma in the inelastic γ-quanta scattering cross sections coincide well with neutron binding energies. As the excitation energy increases the cross section of inelastic scattering is increasing too, and a maximum is formed due to the giant resonance in the photoabsorption cross section, as seen in the case of the reaction (γ, γ') on yttrium.

Because of the continuous nature of bremsstrahlung a yield Y of any photonuclear reaction has the form

$$Y = A \frac{Ni}{M} \int_{0}^{E_e} \sigma(E_\gamma) \Phi(E_e, E_\gamma) \, dE_\gamma, \qquad (8.18)$$

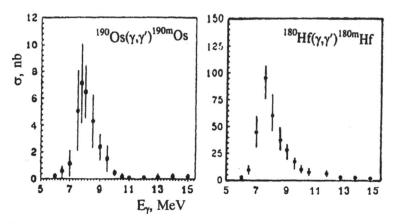

Figure 8.11. Cross section of the reaction (γ, γ') on hafnium and osmium (from [8.4]).

where A is a geometrical factor, N is the total number of nuclei in the target, i is the current of electrons striking the bremsstrahlung target, E_e is the electron energy, $\sigma(E_\gamma)$ is the cross section of the given reaction, $\Phi(E_e, E_\gamma)$ is the bremsstrahlung spectrum, and E_γ is the energy of a photon.

The photonuclear reaction cross section directly determines the yield of the reaction. In a wide sense, the yield of the given reaction Y is defined as the rate of production of the final product under specified bombarding conditions. In the particular case of a photonuclear reaction under bremsstrahlung the yield means the number of secondary particles $(n, p, \alpha, t, ...)$ originated in the unit mass (1 g) of the sample in the unit of time (1 s) per unit of incoming flux of radiation at given electron energy.

All the cited peculiarities of photonuclear reactions are clearly seen in Fig. 8.12, where yields of different photoreactions depending on atomic number of a nuclide, under irradiation by bremsstrahlung from 30 MeV electrons, are shown [8.5].

8.3 Photofission of Heavy Nuclei

For heavy nuclei, fission is one of the decay channels of an excited nucleus. At fairly low excitation energies (to about 30 MeV for photons) a nuclear reaction predominantly passes through the stage of formation of the compound nucleus, and therefore the main properties of outgoing products do not depend on the type of exciting particle. Some differences one can observe only in the near-barrier region which equals $\sim$6 MeV.

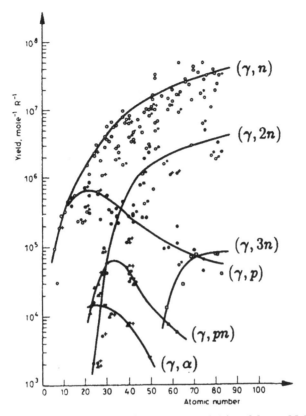

Figure 8.12. Yields of photonuclear reactions initiated by 30 MeV electron bremsstrahlung for nuclides with different atomic numbers (from [8.5]).

As the barrier heights for heavy nuclei equal only about 6 MeV and the fission energy is about 200 MeV, the physical properties of fragments (kinetic energy, mass and charge distributions) in spontaneous and induced fission are practically the same.

Upon the fission of a nucleus into two fragments, these turn out to be excited and the excitation is removed mainly by neutron emission. But the excitation energy and therefore the number of emitted neutrons depends on the mass of the fragment. All the formed fragments possess β- and γ-activity, but really we can only detect activity at the very end of the chain of β-transitions. Most fission fragments are radionuclides with half-lives from a fraction of a second to some years. The time dependence of the integral γ-activity of uranium fragments in different energy intervals and at different irradiation times by bremsstrahlung from 7 MeV electrons is shown in Fig. 8.13. It is seen that γ-activity rapidly decreases with time.

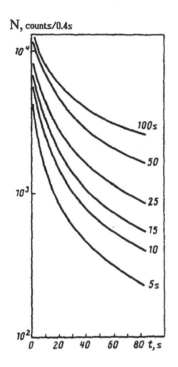

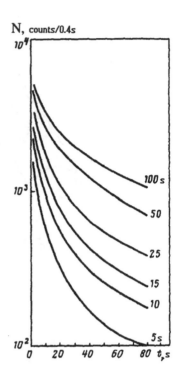

Figure 8.13. Time dependence of induced integral γ-activity of uranium at different times of irradiation. Electron energy $E_e=7$ MeV. *Left* — energy interval 70–122 keV. *Right* — energy interval 270–333 keV (from [8.6]).

In the fission of heavy nuclei, the masses of fragments vary greatly. Fission fragments may have both nearly the same (symmetric fission) and markedly different (asymmetric fission) mass numbers. The most salient feature of mass distributions for nuclei heavier than thorium ($Z = 90$) is their asymmetry: maxima in the mass number ranges $A = 90$ to 110 (light fragments) and $A = 130$ to 145 (heavy fragments). This means that heavy nuclei are divided with the greatest probability into two fragments of unequal masses. At the same time, in the mass distribution of radium one can observe three maxima corresponding both to symmetric and asymmetric fission. The yield of symmetric fragments is increased with excitation energy, but asymmetric fission continues to be dominant in the energy region discussed, as it is seen in Fig. 8.14.

In a number of fragment nuclei, the energy of β-decay turns out to be higher than the binding energy of the neutron in the

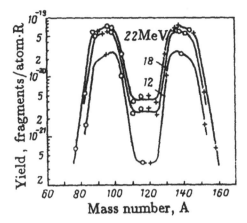

Figure 8.14. Mass distribution of fission fragments in photofission under bremsstrahlung with different endpoints [8.7].

daughter nucleus, resulting in delayed neutron emission. More than 50 emitters of delayed neutrons are known, their decay half-times ranging from 0.2 s to 55 s. The actual neutron emitting nuclides have an immeasurably short half-life, but since they depend for their existence on the β-decay of parent nuclides, they have an apparent half-life equal to the half-life of the β-active parent. As an example, in Fig. 8.15 the chain of radioactive transformations leading to the delayed neutron emitter $^{87}_{36}$Kr is shown.

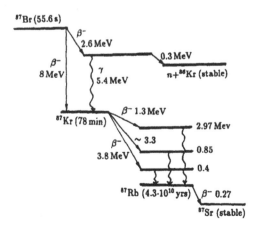

Figure 8.15. The chain of radioactive transformations of the fission fragment ^{87}Br.

The systematic study of the dependence of the number of

prompt neutrons on excitation energy in the photodisintegration of eight actinide nuclei in the energy region 8–13 MeV was made by Caldwell *et al.* [8.8]. The delayed neutron yield greatly varies from nucleus to nucleus, and the difference in the characteristics of the neutrons, emitted in the photofission of various nuclei, may provide a basis for the nondestructive control of nuclear materials. The greatest difference should of course be observed near the barrier, and therefore of special interest are the measurements of the integrated yields of prompt and delayed neutrons due to bremsstrahlung. Such data from [8.9] are presented in Fig. 8.16.

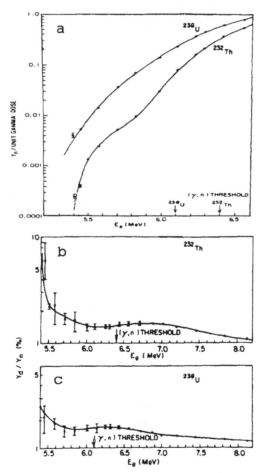

Figure 8.16. The yields of photofission neutrons: *a* – prompt neutrons, *b,c* – bremsstrahlung-averaged delayed neutron fraction for ^{232}Th and ^{238}U (from [8.9]).

Chapter 9

Photoactivation Analysis

The origin and development of activation analysis is a result of combining nuclear physics and radiochemistry. The classical activation analysis method, that is ·the elemental one, or more correctly isotopic analysis, is that the substance under investigation and the standard sample with a known quantity of the determined element are irradiated by a fixed quantity of radiation, and then (immediately or after some time-interval) the induced radioactivity is measured. The existence of an element is identified by the type of emerging particles, their energies and their life-times.

Apparently, the earliest work in photonuclear activation analysis was done by Gorshov and coworkers [9.1], who determined beryllium by prompt neutron emission with a radium source. This work was accomplished in the late 1930's, which makes it contemporary with the experiments of Hevesy and Levi [9.2] and of Seaborg and Livingood [9.3], which were considered the genesis of activation analysis .

The rapid development of applied nuclear physics — nuclear reactors, accelerators, different detectors of radiations, electronics — led also to the extraordinary development of activation analysis, and it is now one of the most sensitive methods of analysing substances.

In spite of the seeming simplicity of activation analysis, the problem of the determination of the content of the element, or a group of elements, in every substance has to be solved practically from the very beginning on the basis of the expected concentration level, the amount of material to be analysed, its physicochemical properties, and the experimental possibilities at the researcher's disposal. Many special investigations are devoted to these problems. We are interested here in potentialities of photoactivation analysis (PAA) and therefore consider below its basic principles and then we shall discuss concrete examples related to different fields and different objects, and consider the problems that have been solved by using different experimental techniques.

9.1 Peculiarities of Photoactivation Analysis

It is necessary first to note that there are a lot of elements that are slightly activated by neutrons (neutron activation analysis — NNA) and are determined well by photonuclear methods. The following elements belong to this group: Be, Mg, P, Ca, S, Ti, Se, Zr, Mo, Cd, Tl, Pb, and Bi. In spite of the fact that NAA provides the higher sensitivity (the detection limits in NAA are 2–3 orders of magnitude higher than in PAA), it is preferable to use PAA when it is necessary to obtain high expressivity and productivity. This is explained by the fact that neutron-deficient nuclei are formed in the (γ, n)-reaction and they, as a rule, possess shorter half-lives than radionuclides of the same element under neutron irradiation. The following elements can illustrate this statement (the corresponding half-lives under neutron and photon activation are given in brackets): K (12.5 h; 7.7 min), Na (15 h; 23 s), Mg (9.5 min; 11 s), Sc (84 d; 3.9 h), Fe (45 d; 8.9 min), Zn (31 d; 4.4 min), Se (120 d; 17.5 s), Br (1.48 d; 5.0 s), Sr (64 d; 16.1 s), Zr (31 d; 4.4 min), Ba (83 min; 2.5 min), Ta (115 d; 8.1 h), W (23.8 h; 5.3 s), Ir (74 d; 4.5 s), Au (2.7 d; 7.2 s), Hg (46.6 d; 4.3 min).

The matrix composition is also an essential point in the choice of analysing method. In particular, analysis of ores and rocks are favourable objects for PAA, and this is due to the rather high threshold energy of basic rock elements — Na, K, Si, O, Al. On the other hand, these macroelements of ores and rocks have a large neutron activation cross section that leads to considerable background in determination of ore-forming elements. The second, but not the last, advantage of PAA application in geology is the large penetration length of primary radiation — high-energy photons — that makes it possible to analyse massive samples, providing high representativity of the analysis. The use of photonuclear method is also preferable in the case of determination of the content of elements that are extremely inhomogeneously distributed over the sample. The most well known case is the analysis of gold content in different geological specimens.

The photonuclear method also has an advantage in the analysis of biological objects. Sodium, potassium, and chlorine, highly activated by neutrons, are involved in the main macroelements of biological tissue, and they interfere in the determination of the microelement content forcing analysts to use radiochemistry.

In modern activation analysis laboratories, computers, particularly personal computers (PC's), are commonly used to process the large amounts of spectra generated, including functions such as storing spectra, transferring spectra to other computers, computations of activity and concentrations, and can also be used to treat all the information obtained.

Similarly as with instrumental NAA, as important aid with PAA in the interpretation of measures gamma-ray spectra measured (in radionuclide identification) is represented by tables of photonuclear reaction products, comprising precise data on the energy of photons emitted by radionuclides, the intensities of gamma lines, the half-lives of decays, and the nuclear reactions with a proper reaction threshold etc. The most complete data, presented in a very suitable form for practical use, was published by Řanda and Kreisinger [9.4].

9.1.1 Transport System.

The samples for irradiation are commonly placed in aluminium cartridges which are then conveyed by a pneumatic tube from the analytical laboratory to a position immediately behind the target.

The requirements of industrial production processes are best met by instrumental activation analysis using short-lived isotopes; in many cases, the sensitivity and selectivity are superior to those obtained in analysis using elements with long radioactive half-lives. Another important requirement of activation analysis is that the probes be reasonably easy to handle (of volume 50-$100 \, cm^3$) and that the radiation field is uniform.

For laboratory investigations it is also desirable to have a simple and reliable transport system to transfer samples for irradiation and back to the counting system. We consider here a simple tube conveyer for quick activation analysis of probes [9.5] that can be adapted easily for a wide range of problems to be solved.

As a rule, in conveyer tubes, either metal tubes, which are very difficult to bend without altering their cross sections, or polyethylene tubes of small diameter, are used as the transport channels. Polyethylene tubes of larger diameter cannot be used directly because their cross sections are invariably elliptical. However, this defect is easily removed by taking a spring wire (diameter $3 \, mm$) and winding it in spirals of internal diameter less than the external diameter of the tube, after which the spirals are tightened against the outside of the tube. In this way the authors designed a conveyer channel of length 30 m, inner diameter 45 mm, and wall thickness 2.5 mm. The containers were polyethylene boxes of length 50 mm, diameter 40 mm, and wall thickness 1 mm. Air pressure from an ordinary vacuum cleaner moved the containers along the polyethylene tube in both directions; the time required to transport the boxes is 3–4 seconds.

Figure 9.1 displays a diagram of the entire conveyer. The vacuum cleaner *1* produces excess pressure or a partial vacuum depending on how the switching device *2* is set. The switch *2* is designed as a sliding contact and is connected with the loading chamber *4* by means of a flexible hose *3*. The loading chamber consists of two telescopic brass tubes; the inner tube is twice the length of the outer one and contains a rectangular window into which the container with the probe is inserted. The window is closed by rotating the outer tube along its axis.

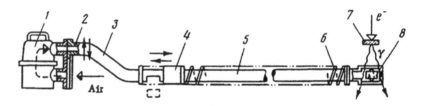

Figure 9.1. Diagram of pneumatic transport device: 1 – vacuum cleaner, 2 – pressure reverser, 3 – flexible hose, 4 – loading chamber, 5 – polyethylene tube, 6 – steel spiral, 7 – bremsstrahlung target, 8 – radiation chamber.

The radiation chamber *8* consists of an outer thin-walled aluminium tube and an inner cylinder which rotates freely about it. The inner cylinder has tangentially directed channels to allow passage of air and a drainage hole at the end. When the container hits the cylinder, the force of the air passing through causes the container to begin to rotate together with the cylinder at ~ 5 rev/sec. The container is induced to rotate by an externally mounted hermetically sealed contact which is activated when the small permanent magnet mounted on the inner cylinder passes by.

For intrachamber irradiation at the microtron a special system for sample rotation was developed that allows (in spite of the inhomogeneity of radiation intensity) to obtain more uniform activation of the sample [9.6]. Moving the radiation chamber in the chamber along the common diameter of the electron orbits, one can change electron energy in a simple way. Intrachamber irradiation can, of course, be used for a short irradiation time not to induce high level of radioactivity in the accelerator.

Probes can be delivered from the loading chamber to the detector either by attaching a flexible stocking or by mounting the loading chamber directly ahead of the detector.

9.1.2   Optimal Size of a Sample in PAA

One of the main advantages of PAA is its large penetrability. Two very essential consequences follow from this property for applications:

(i)   the total volume of the sample can be analysed;
(ii)   the mass of the sample analysed can be rather large, which is, for instance, extremely necessary to provide representativity in the case of examining of content of rocks and ores.

Consider the limitations of the geometric sizes of the sample to be analysed. As a rule, analysed samples are irradiated by bremsstrahlung with the maximum energy 10–20 MeV, and the induced activity is measured by a scintillation spectrometer. The gamma-quantum energies of activation products are usually in the range 0.1–1.0 MeV. Therefore, although the primary penetrates rather deeply into the sample, the secondary radiation from inner regions, possessing considerably less energy, is absorbed. It is obvious, that it is difficult to judge the accuracy and representativity of the analysis in such situations.

Due to these features of photonuclear reactions, for the effective use of primary and secondary γ-quanta the geometric sizes of the sample have not to be the same during the irradiation and measuring of the induced activity. It is rather easy to satisfy these demands. The analysed sample has to be chosen in the form of a cylinder (or close to cylindrical shape) and the dimension h along its axis is ruled by the absorption length of activating γ-radiation in the material of the sample, and the radius of the cylinder is determined by the absorption length of induced activity (see Fig. 9.2).

Secondary radiation is registered by a ring crystal with inner size equalling the size of the sample. So, the effective geometric sizes of the sample during irradiation and counting are different. The dependence of the size of the sample and the detector on the energy of the irradiating and registered γ-radiation allow in any given case of PAA to choose their proper optimal sizes. At the same time, these characteristics determine the maximum mass of the probe which can be analysed.

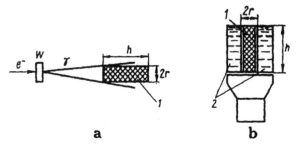

a   **b**

Figure 9.2. Schematic drawing of irradiation of the sample (*a*) and measuring of the induced γ-activity by a ring detector (*b*): e⁻ — electron beam, W — bremsstrahlung target, γ — bremsstrahlung, 1 — analysed sample, 2 — scintillator.

To verify these optimal conditions experiments with gold ores (gold content 50 g/t) have been carried out [9.7]. One part of the analysed ore (~ 200 g) was packed in a 'standard' cylindrical capsule of diameter 60 mm and height 60 mm, and another into an 'optimal' cylindrical capsule with diameter 18 mm and height 35 mm. The capsules were irradiated by bremsstrahlung from 9 MeV-electrons. The irradiation time was 18 s, cooling time 3 s, and activity measuring time 18 s. The induced γ-activity was measured by a NaI(Tl)-scintillation crystal 70 × 70 mm, and the 'optimal' probe was measured by similar NaI(Tl)-crystal having a well 18×35 mm.

Owing to (γ, γ')-reaction in gold, γ-quanta with energy 277 keV are originated in the sample, and this energy region of the spectrum is shown in Fig. 9.3. The data presented show that in spite of the decrease in the mass of the analysed sample in the 'optimal' capsule by 4 times, the useful spectrometric information increases by about 5 times, confirming the arguments given above. Actually, experiments and calculation show that counting efficiency in the case of 'common' geometry is about 20%, and this means that in 4π-geometry of detecting system for 'common' size sample, the maximum increase of the counting rate is 5 times, but for the 'optimal' size sample the efficiency is increased by 20 times.

9.1.3   Recoil Nuclei

As a result of nuclear reaction, such as (γ, n) or (γ, p), the nucleus escape a nucleon and, at the same time, the nucleus itself acquires a recoil energy that can be enough to leave a sample under irradiation. For example, as a result of (γ, n)-reaction a

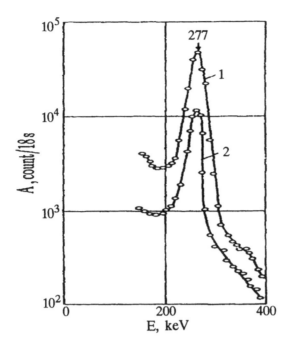

Figure 9.3. Gamma-spectra of induced activity in gold-content ore for 'common' (1) and 'optimal' size (2) samples.

fluorine nucleus acquires an energy $0.05\,\mathrm{MeV/aum}$ and its range in teflon — $(CF_2)_n$ — is approximately equal to $1\,\mu\mathrm{m}$. This feature can be successfully used for preparing radiogenic isotopic sources. Such specimens with a known content of radioisotope are, in particular, necessary for experimental verification of the process of radiochemical extraction used in a given activation analysis. It is essential that the method of recoil nuclei permits to insert labelled nuclei in those chemical forms as they are stabilized during irradiation.

Let us use fluorine to illustrate such a technique. The method of obtaining a specimen with implanted recoil fluorine nuclei implies that thin foils of analysed material, stacked with the teflon films, are irradiated by bremsstrahlung. The number of recoil nuclei, implanted into different materials, differs because of backscattering, and this effect reaches a significant value for materials of high Z. In principle, one can calculate this number, but a lot of difficulties arise due to the need to take into account energy losses and multiple scattering of recoil nuclei during their motion in a substance, and inherent energy and angular distribution of recoil nuclei.

Experimental determination of a quantity of recoil fluorine nuclei implanted into different materials from teflon was performed by Murashov [9.8]. For these experiments were chosen materials which under activation do not form positron emitters with $T_{1/2}$ commensurable with $T_{1/2}$ of ^{18}F – aluminium, vanadium, niobium, tungsten. Foils or thin plates of these materials stacked with teflon films are irradiated by bremsstrahlung with maximum energy 30 MeV. The number of recoil nuclei implanted into the samples were measured after decay of radionuclide ^{12}C that is also formed in teflon. The contribution from inherent induced activity was taken into account by measuring the activity of materials irradiated without teflon films under the same conditions. The results obtained are shown in Fig. 9.4.

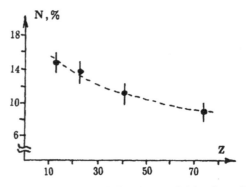

Figure 9.4. The quantity of recoil fluorine nuclei implanted into different materials (percent of the number of ^{18}F formed in a 1 μm-layer of a teflon) (from [9.8]).

We see that the method of recoil nuclei can effectively provide a sample with a known quantity of radioactive nuclei that can be used for testing any radiochemical or physicochemical separation technique, as well as for preparation of labelled specimens.

9.2 Photoactivation Analysis of High-Purity Materials

In the mid 60's the need to determine nonmetal impurities in metals with sensitivity up to $10^{-4} - 10^{-6}\%$ mass arose due to the wide use of high-purity materials in science and technology — semiconductors, superconductors, alkaline and alkaline-earth metals, single crystals of hard-melting metals, multicomponent compounds. This problem was always rather difficult and was practically an intractable one, due to the existence of impurities on the metal surface and gas release from apparatus and reactives during the analysis. In particular, the problem of the determination of gaseous impurities (C, N, O) in pure substances often attracts attention as it is rather difficult to extract them from the bulk as they naturally influence the physical properties of substances to be investigated [9.9, 9.10]. The possibility of determining gaseous impurities gives way to control process of high-purity substance production, and, on the other hand, one may investigate the influence of gaseous impurities on the physical properties of substances such as cross sections of phonon and electron scattering, on electron energy spectra and so on. One can use charged particles for this problem but only a small surface layer can be analysed due to their short ranges.

The (γ, n) activation products of carbon, nitrogen, and oxygen all decay exclusively by positron emission, and are detected by the associated annihilation radiation. At low levels, unless the sample is quite simple and its gamma activation products do not yield positron emitters, it is frequently necessary to perform a chemical separation. Nevertheless, in some cases light elements can be determined non-destructively as in the case of calcium and boron. The photonuclear products of these elements do not interfere with the counting of the oxygen activity, and, if other impurities are present only at very low levels, sensitivities of 1 to 0.01 ppm are obtained.

9.2.1 Nondestructive Photoactivation Analysis of Impurities

As it was mentioned above, in some high-purity materials we can analyse gaseous and even metal impurities nondestructively, as is illustrated by the following examples [9.11].

9.2.1.1 *Oxygen in Very Pure Sodium*

By activating sodium in gamma rays, one obtains principally two active isotopes according to the following reactions:

$$^{23}\text{Na}(\gamma, n)^{22}\text{Na} \qquad T_{1/2} = 2.58 \,\text{y}, \qquad E_{\text{thr}} = 12.5 \,\text{MeV}$$

$$^{23}\text{Na}(\gamma, \alpha n)^{18}\text{F} \qquad T_{1/2} = 1.7 \,\text{hr}, \qquad E_{\text{thr}} = 21 \,\text{MeV} \ .$$

The half-life of the first is such that the specific induced activities are very weak. With respect to fluorine-18, experience shows that its activity diminishes very significantly if the energy of the electron beam is lower than 25 MeV. Thus, the nondestructive determination of oxygen proves to be possible by irradiating with energies lower than 25 MeV.

To determine very small quantities, it is moreover necessary that the sodium be very pure and contain, above all, little potassium. This element is activated very strongly and yields potassium-38 with a half-life of 7.7 min according to the reaction

$$^{39}\text{K}(\gamma, n)^{38}\text{K} \quad \text{with the threshold} \quad 13.1 \text{ MeV} .$$

For example, to determine the oxygen content of the order of parts per million, it is necessary that the potassium content be lower than 10 ppm. An example of the decay curve for the determination is shown in Fig. 9.5.

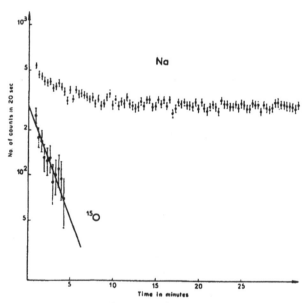

Figure 9.5. Decay curve for a nondestructive determination of oxygen in very pure sodium containing approximately 10 ppm potassium; $E_{\text{max}}=25$ MeV, $I_{\text{av}}=25$ μA, $O_2=10$ ppm (from [9.11]).

An anticoincidence method exists for reducing this interference due to potassium originated by Holm [9.12], and the estimation shows that one can gain a factor of at least 5. For this type of analysis, a sodium sample is sealed in an airtight ampoule of quartz, whose bottom and lateral sides are covered with a thin foil of aluminum. After a 5-min irradiation, the ampoule is broken and the aluminum foil adhering to the sodium is stripped, which is a first mechanical cleaning. The sample is then rapidly etched in a solution of ethyl alcohol with 10% water, dried, and counted.

Oxygen in calcium can be determined in the same way. In this case, it is necessary to irradiate at still lower energy, of the order of 20–22 MeV. Calcium in effect yields potassium-38 according to the reaction:

$$^{40}Ca(\gamma, n)^{48}K \quad \text{with the threshold} \quad 19.2\,MeV\ .$$

Likewise, Persiani *et al.* [9.13] carried out the non-destructive determination of oxygen and carbon in cesium. They showed that at 28 MeV and higher the matrix activated more and more. The amounts determined are of the order of a few parts per million for the two impurities.

In these materials one can naturally determine nitrogen and fluorine in a nondestructive manner.

9.2.1.2 *Nondestructive Determination of Carbon, Nitrogen, Oxygen, and Fluorine in Very High-Purity Beryllium*

Beryllium practically does not activate. Thus by activation in gamma rays one produces mainly the following two reactions: $^9Be(\gamma, n)^8Be$, which instantaneously disintegrates into two alpha-particles, and $^9Be(\gamma, 2n)^7Be$, with a half-life of 53.4 days whose threshold is at 20.6 MeV. Even for irradiation energies of 30 MeV, the induced activity remains weak, with the result that in the non-destructive determinations, the activity due to ^7Be which is a gamma emitter, $E_\gamma = 0.48\,MeV$, is buried in the background noise of the coincidence counters.

The metal, therefore, is an ideal matrix for nondestructive analysis. The example of classic decay curves relative to the determinations of oxygen and nitrogen in a beryllium sample is shown in Fig.9.6.

9.2.1.3 *Determination of Metallic Elements*

Experience shows that PAA is interesting not only for the determination of the light elements ·but also for the metallic impurities. Schweikert and Albert [9.14] have shown that one can, for instance, determine zirconium in hafnium and silver in bismuth in a nondestructive way, and can also determine titanium in iron

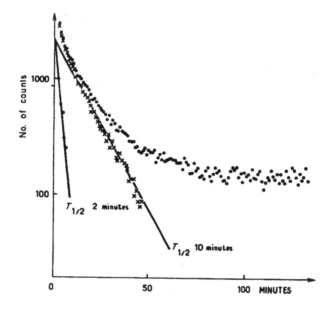

Figure 9.6. Decay curve for nondestructive determination of oxygen and nitrogen in a beryllium sample. The duration of the irradiation is 2 min at 30 MeV; due to the high nitrogen content, the activity of ^{62}Cu ($T_{1/2}=9.8$ min) is negligible; O=10 ppm, N=140 ppm, Cu=6.4 ppm (from [9.11]).

and nickel in copper with a very high sensitivity. For the latter determination, Cuypers has shown [9.15] that the determination of nickel in copper by activation with thermal neutrons is impossible for amounts smaller than a few parts per million because of the nuclear reactions on copper. In this very particular case, activation with gamma rays therefore permits the removal of this difficulty.

9.3 Determination of Nonmetal Impurities in High-Purity Materials by PAA Combined with Radiochemical Extraction

The gamma-activation method possesses two main advantages:

1. Owing to the large penetrability of high-energy γ-quanta the element to be investigated is activated practically uniformly over the whole volume of the sample that provides the representativity of the sample and correct monitoring of the bremsstrahlung flux passing through the sample; samples of any shape and up to 0.5 g mass can be used.

2. The half-lives of the obtained radionuclides ^{15}O (2.1 min), ^{11}C (20 min) and ^{13}N (9.9 min) are large enough to eliminate by etching the surface of a radiated sample before the chemical separation of radionuclides; in high-purity metals the amount of surface oxygen may be much more than in the bulk.

Consider now, how these advantages have been realized in photoactivation analysis of gaseous impurities in various substances performed at the 30-MeV microtron of the P.L. Kapitza Institute for Physical Problems (Moscow) during the last 20 years.

9.3.1 Irradiation Set-up and Detection System

The diagram of the experimental lay-out is shown in Fig. 9.7. The samples to be analysed are irradiated by the bremsstrahlung produced at the internal or external target of the 30 MeV-microtron. Samples up to 9 mm in diameter with a mass of 0.5 to 4 g are placed in an aluminium or polyethylene rabbit between two small monitor platelets. At the external irradiation the rabbit was sent to a 5 mm tungsten target by a pneumatic tube transfer system (the transport time was about 6 s). The irradiation lasted approximately three half-lives of the corresponding radionuclide and then the sample transferred back was chemically etched to dissolve up to 0.1 mm of metal over its surface, using, for example, nitric acid for copper, and fuming perchloric acid for steel. After etching the irradiated sample is weighed and a radiochemical extraction of the radionuclide is performed.

The sensitivity of the method can be estimated from the specific activity of the radionuclide released on the sample after the end of irradiation and the activity of the corresponding monitors.

The impurity content is calculated from the ratio of the specific activities of a radionuclide separated from the sample and measured in monitors.

A gamma-gamma coincidence spectrometer is used to count the positrons formed by the disintegration of the radionuclides (^{11}C, ^{13}N, ^{15}O). Two 150 mm in diameter per 100 mm NaI(Tl)-crystals are used for detection of annihilation gamma-quanta. The background count of the spectrometer is 2 cpmin, and the efficiency of positron detection equals 13%. The radiochemical purity of the separated radionuclides is checked by the half-life of the activity detected.

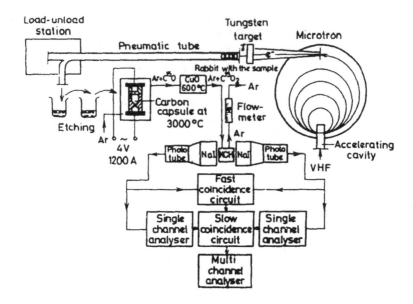

Figure 9.7. Complete experimental arrangement for oxygen determination by high-energy γ-quanta irradiation at the IPP microtron [9.16].

9.3.2 Methods of Radiochemical Extraction of Radionuclides

9.3.2.1 Oxygen

For oxygen determination in metals a reductive melting method in an inert gas stream, using graphite capsules, for the rapid radiochemical separation of ^{15}O has been developed [9.16]. This method has some advantages over vacuum melting in an inert gas stream, particularly if the operation is carried out in capsules, as extraction losses of carbon monoxide due to sorption on sublimates are substantially lower. Further, pulse heating to high temperatures allows oxygen extraction in the course of seconds. The time required to heat the capsules with the sample to 2800°C is 10–20 seconds. At such high temperatures, rapid and complete reduction of the oxides occurs, accompanied by the formation of $C^{15}O$. Carbon monoxide is carried away from the reaction chamber in an argon stream, passes through a CuO-filled oxidation furnace at 600°C, where it is oxidized to $C^{15}O_2$ which is subsequently absorbed in a trap filled with KOH powder. The trap is placed between scintillator crystals from measuring the induced activity. The time required for the CO_2 transfer was found to be 50 seconds.

In order to ensure that oxygen release in the melt proceeds at the optimum condition for the pulse heating process, and that the degree of release does not depend on the oxygen content in the sample, oxidized nonirradiated metal was added to the sample in the capsule. The oxygen content of the nonirradiated metal served as a carrier. In analysis of copper and indium 0.5–0.8 g of copper powder containing 0.8% oxygen was added, while in iron and steel analyses, oxidized steel chippings were used. In addition, when copper and indium were analysed, a nickel bath in 1:1 proportion to the weight of the metal was applied in order to improve the extraction of oxygen by increasing the equilibrium concentration of carbon, since carbon is poorly soluble in copper and indium.

Figure 9.8 illustrates the decay curve of ^{15}O, separated from a steel sample. The sample measured contained about 1.8 ppm of oxygen. The whole separation procedure was accomplished in 2.5 min after the end of irradiation. The measuring of the activity lasted 15 min. Then the activity of ^{15}O in boron trioxide monitors was measured under the same counting conditions. On the same Fig. 9.8 (insert) the decay curve of ^{15}O separated from a copper sample with an oxygen content of 0.36 ppm is shown. Even in this case the identification of ^{15}O by half-life measurements is still possible.

9.3.2.2 Carbon

A sample of a mass ~0.5 g together with two graphite standards (2.6 mg) is irradiated during 15 min. After irradiation the surface layer of a width of 100 μm is removed by etching in a hot mixture of HCl-HNO$_3$ (3:1) during 60 seconds. Then the sample is weighed and is melted with a carrier (10 mg of graphite) in a waterless alkaline oxidative melt of 3 g KOH and 3 g KNO$_3$ at 600° C in a nickel crucible pumped by a waterstream. Gases released during alloying are oxidized on copper oxide and then are absorbed by 12M KOH (10 ml) to control carbon losses during the process of alloying. The nickel crucible with the melt is removed from the quartz glass, is cooled and dissolved in 50 ml of a warm water. The solution obtained is transferred into a round 500 ml Pyrex flask with a back refrigerator, 10 ml of concentrated HNO$_3$ is added and all together is boiled for 3 min with air flowing through the flask and a trap with 12 M KOH during 10 min. The alkali solution is then transferred into a flat polystyrene cassette of a volume of 10 ml and its activity is measured by

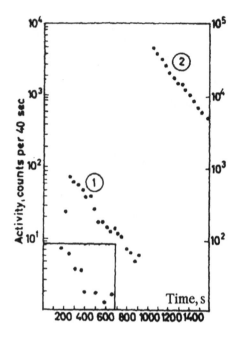

Figure 9.8. Typical decay curves of ^{15}O: 1 – separated from 2 g steel sample, contained 1.8 ppm of oxygen; 2 – measured in monitors ($\sim$ 200 mg of boron trioxide). *Insert*: separated from 1.6 g copper sample, contained 0.36 ppm of oxygen [9.16].

a (γ-γ)-spectrometer for 20 min, and then the activity of the graphite monitors is measured.

According to the second method the sample is alloyed with 6 g of annealed KNO_3 and 10 mg of graphite at 600°C in a nickel crucible. The flask with a crucible is blown by air for 15 min and then the alkali solution is transferred into a cuvette to measure its activity.

The radiochemical yield of carbon was equal to 80-95% and the detection limit was about $3 \cdot 10^{-6}$ mass %.

9.3.2.3 Nitrogen

For nitrogen determination the technique of high-temperature extraction at 2000°C with subsequent absorption of removed molecular nitrogen by a titanium sponge was developed, and an alumina nitrite is used as a monitor. After etching, the sample with 5 mg of alumina nitrite and 0.3 g of electrolytic copper is fed into a graphite capsule and then is placed between the electrodes of the extraction set-up. After 2 min argon blowing off (200 ml/min) the capsule is heated up to 2000°C with two 1000 A and 30 s duration current pulses from the transformer. After 15 min a quartz flask with titanium ($10 \, cm^3$) is cooled and its activity is measuring during 20 min. The detection limit is equal to $\sim 5 \cdot 10^{-6}$ mass %.

9.4 PAA of Gaseous Impurities in Metals

The methods described above for determining of oxygen, nitrogen and carbon were applied to solve various problems such as for high-purity metal analysis and testing of methods of their production (Fe, Nb, Cu, In, Sn, Si, Al, Na, Li, W, Mo); for investigation of the efficiency of In, Fe, Mo, Si purification from gaseous impurities; for studying of the spatial distribution of oxygen, nitrogen and carbon spatial in In, Nb, Mo, Si after different methods of purification, in particular after zone melting; for testing the purity of single crystals (W, Si) after growing them using different techniques.

Some samples were analysed without preliminary surface etching. The oxygen content found in steel samples where the surface layer was removed by etching, corresponds approximately to the lowest value in the quality certificate, while the oxygen percentage found in samples without preliminary etching exceeds the highest value in the certificate. The effect of oxygen absorbed on the surface is particularly important in copper and tin analyses. The true oxygen content in the bulk of the sample is lower by approximately one order of magnitude than the value found by vacuum melting and activation analysis without preliminary etching of the surface [9.17, 9.18].

The results of PAA of small bulk oxygen content in Nb, Fe, and In (Tables 9.1–9.3) were compared with the results obtained by the reduction fusion method in a vacuum with ionic etching of the surface, with internal friction technique, and with the data calculated from the annealing conditions during low-oxygen content production of the niobium and iron samples (less then 10^{-5} mass %) [9.19, 9.20]. The PAA results, $(1\text{-}3) \cdot 10^{-5}$%mass, coincided with those calculated by thermodynamical and kinetic approaches and with the reduction fusion method. It has been shown that there were 2-6 μg of oxygen on the surface of high-purity iron samples. It also followed from the data obtained that a niobium sample can be produced with an oxygen content less then $2 \cdot 10^{-5}$%mass, and this conclusion was also confirmed by the internal friction measurements [9.19].

Table 9.1. The results of oxygen determination in some SRM samples of iron

Sample	Oxygen content, ppm (quality sertificate)	Determined by PAA, ppm
SRM 98 ppm	98	90, 130, 93
SRM Steel-45 (USSR)	32	31, 28.5, 30
SRM 1094 (NBS, USA)	2.5 – 7.5, mean value 4.5	2.9, 2.2, 2.2
SRM 1090 (NBS,USA)	484±14	479, 481, 488, 442

Table 9.2. The results of oxygen determination in niobium, prepared in ZFW, Dresden

Sample	Introduced oxygen, ppm	Reductive fusion, ppm	Internal friction, ppm	PAA, ppm
Nb-160 3.1	60	51	82	50, 58
Nb-160 3.2	100	126	126	150, 110, 120
Nb-161 3.1	200	253	210	150, 140
Nb-161 3.2	25	45	45	65, 77
Nb-170 3.2	–	–	–	14, 14, 12

Table 9.3. Determination of oxygen in pure indium

Sample	Reductive melting in argon, ppm	PAA, ppm
In-1	4 ± 3	2.3±0.7
In-2	–	1.0, 0.8
In-3 from peripheral of the bar	–	2.4 ± 0.4
In-3 from central bar	–	0.7 ± 0.1

The advantages of PAA have been fully exploited in analyses of chemically active alkaline and alkaline-earth metals, in which nonmetal impurities on a surface are considerably more than in the bulk. In particular, by using bremsstrahlung from the 30 MeV-microtron it was managed to determine an oxygen content at the level $2 \cdot 10^{-5}\%$ mass in sodium and lithium [9.21]. In these cases a special method was developed for oxygen extraction from the samples based on a bulk oxygen exchange with the oxygen in water during sample dissolving after its etching (0.4 mm of the surface layer was removed). Subsequent distillation of the water over 2 minutes allows the extraction of ^{15}O from other radionuclides formed in the matrix. The reproducibility of the results at oxygen contents $10^{-2} - 10^{-3}$ mass % was about 0.1.

The method of nitrogen determination in sodium and lithium was also developed based on ammonia distillation with a detection limit for nitrogen $2 \cdot 10^{-5}$ % mass [9.22, 9.23].

Consider now the methods of carbon content determination in high-purity ($< 10^{-4}$ %) Fe, Mo and W. For radiochemical extraction of ^{11}C from iron, the method of metal oxidation in an oxidizing melt with Pb_3O_4 (87.5%) and B_2O_3 (12.5%) during inductive high-frequency heating was used [9.24]. Therefore, using this technique it was shown for instance that the carbon content in iron annealed in hydrogen or cleaned by zone melting was $3 \cdot 10^{-5}$ %, but this value can be decreased by applying both these techniques up to $7 \cdot 10^{-6}$ %. The results also showed that molybdenum with a carbon content $(3-7) \times 10^{-5}$ % can be produced with annealing in an oxygen atmosphere at pressure 2.7×10^{-3} Pa, and temperature $1550 - 1650°C$ during $2 - 6$ hours [9.25].

For the radiochemical determination of carbon in a W-single crystal the method of radiochemical extraction of CO_2 was developed based on the oxidation of tungsten in an alkaline melt with NaOH (50%) and $NaNO_3$ (50%) with a detection limit equalling $1.5 \cdot 10^{-6}$ % mass. The method was checked by analysing of synthetic tungsten sample with a carbon content of 3×10^{-2} %. Tungsten single crystals obtained by remelting in an argon plasma of metallic samples made from their oxides, chlorides and fluorides, were analysed, and it was founded that the carbon content was no less than a few tens of nanograms [9.26].

There is little data concerning the influence of nitrogen on different properties of metals and semiconductors, and therefore we have developed the special methods for analysing the nitrogen content in Al and Si. The PAA of nitrogen determination in Al [9.27] is based on dissolving a metal in 6M KOH and a distilling ammonia and molecular nitrogen. Ammonia was oxidized in CuO to obtain molecular nitrogen, which then was absorbed from an argon flow by a titanium sponge at $900°C$. By applying this technique it was, in particular, obtained that in aluminium produced by single electrolysis, the nitrogen content is comparable with the quantity of oxygen (6×10^{-4} %).

In silicon a carbon and nitrogen were analysed simultaneously [9.28], by melting silicon with KOH. From the melt obtained nitrogen is extracted in the form of ammonia and carbon in the form of methane. The experiments showed that the detection limits are $2 \cdot 10^{-5}\%$ mass and $3 \times 10^{-6}\%$ mass for nitrogen and carbon respectively.

9.5 Alloys and Salts

The main components and the doping impurities of alloys are currently determined by different chemical or physico-chemical methods. There exist extremely few procedures adopted for using the activation method in the analysis of alloys, which is probably accounts for its traditionally, preferential use as a high sensitivity method for the analysis of high-purity substances and mineral raw materials. On the other hand, using semiconductor gamma-spectrometry makes it possible not only to perform high-sensitive activation analysis but also to determine rather high concentrations of elements with high sensitivity.

9.5.1 Impurities in Salts

Impurities in salts are present as H_2O, anions OH^-, NO_3^-, CO_3^{2-}, NH_4^+, organic compounds and carbon. As far as chemical forms of stabilization of recoil atoms ^{11}C, ^{13}N, ^{15}O made during the (γ, n)-reaction are not known, the radiochemical method has to provide their isolation regardless of stabilization forms.

Let us consider the determination of $1-10$ ppm of nitrates in thallium bromide. After irradiation samples were etched by solution H_2SO_4 to remove a 0.3 mm layer. Oxygen and nitrogen were simultaneously extracted in a flow of argon by heating a covered graphite capsule, in which the sample was placed up to $2100-2200°C$ for 30 seconds. The gases then passed through $BaCO_3$ at $550°C$ for separation from bromine with decontamination factor of 10^6, CuO and two absorbers with solid KOH and a titanium sponge at $900°C$ (Fig. 9.9).

The separation time from the end of irradiation till the beginning of measurement for oxygen is 5 min, and for nitrogen — 10 min. From the activities of $^{15}O(^{13}N)$ samples and monitors B_2O_3 (AlN) were calculated their masses and ratio in TlBr. The distribution of nitrates along the ingot of TlBr doped by TlNO$_3$ was investigated. The results have shown that the stoichiometry of the nitrate ion remains in the range $1-20$ ppm (Fig. 9.10).

Oxidation of carbon in solutions of metals and alloys was typically performed using iodates and periodates, whereas oxidation of organic compounds was performed in acidic solutions

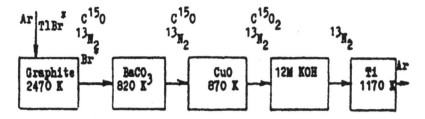

Figure 9.9. Simultaneous radiochemical separation of ^{15}O and ^{13}N from TlBr.

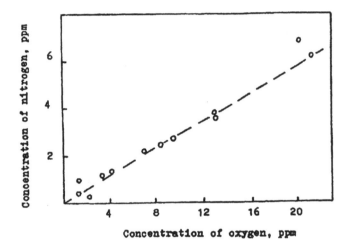

Figure 9.10. The stoichiometry of nitrate ion containing in the ingot TlBr.

[9.29, 9.30, 9.31]. For instance, baths with H_2SO_4-KIO_4 were used for determination of carbon in metals by photon activation [9.30]. The Van Slyke's technique was used for the determination of carbon in thallium halides and KCl, which is based on oxidization of carbon while salts are dissolved in hot concentrated H_2SO_4-H_3PO_4 (2:1) in the presence of KIO_3-$K_2Cr_2O_7$ (2:1), the reagents and the apparatus are described in [9.32]. Tables 9.4 and 9.5 summarize the results of carbon determination in TlBr-TlI, salt and optical crystals of KCl with a detection limit 0.3 ppm.

Melting with KOH was used earlier for the determination of nitrogen and carbon in silicon by photon activation [9.32], and then it was used for radiochemical separation of ^{13}N and ^{11}C from aluminium alloys. After irradiation the samples were etched to remove a 0.2 mm layer. When the alloy is dissolved in an alkaline bath at $550°C$, nitrogen and carbon are reduced by monoatomic

Table 9.4. Carbon determination along the ingot of TlBr-TlI.

Sample	1	2	3	4	5	6
Carbon, ppm	0.50	0.52	0.68	0.58	0.7	0.66

Table 9.5. Carbon determination in KCl.

Sample	Salt	Melting salt	Granules	Crystals
Carbon, ppm	60	130	8.5	1.0-1.7

hydrogen and extracted as $^{13}NH_4$ and $^{11}CH_4$. Then $^{13}NH_4$ is distilled with steam in an argon stream and absorbed in H_2SO_4. Methane is oxidized to $^{11}CO_2$ on CuO at 800°C and absorbed in 12 M KOH solution (Fig. 9.11).

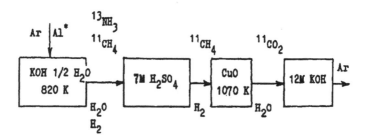

Figure 9.11. Simultaneous radiochemical separation of ^{13}N and ^{11}C from an Al-Mg-Li alloy.

Table 9.6 shows an example of the results of nitrogen and carbon determination in alloys with the detection limit 0.2–0.3 ppm.

The investigation of nitrogen solubility in silicates of various origins has been recognized as an important geochemical problem. Because of the wide variation of nitrogen solubility the analytical method must work equally well from ultralow $(10^{-7}\%)$ up to relatively high levels $(10^{-2}\%)$ of nitrogen concentration. Radiochemical photon activation analysis by the $^{14}N(\gamma, n)^{13}N$ reaction is a promising enough method allowing the removal of surface contaminations and analysis of bulk samples of any geometrical shape with acceptable sensitivity and accuracy.

The special experiments show the absence of nuclear interferences due to the reaction $^{16}O(\gamma, t)^{13}N$ if the energy of

Table 9.6. Nitrogen and carbon contents in Al-Mg-Li alloy.

Sample	Nitrogen, ppm	Carbon, ppm
A	180, 160	130, 120
B	720, 1000	15, 20
C	2.4, 2, 4	3.2, 2.9
D	2.4, 2.0	3.4, 3.7

bremsstrahlung radiation is lower than the threshold of the above reaction, which equals 26 MeV.

After irradiation and etching of the sample's surface the radionuclide ^{13}N is isolated by means of high temperature (1800°C) extraction. The additional radiochemical purification is done by absorption of ^{13}N with a titanium sponge at the temperature 1100°C. The detection limit for nitrogen content is $10^{-2} - 10^{-3}$ mass% in a 50 mg specimen.

The method developed has been applied for determination of nitrogen solubility in the systems N_2-basalt melt and N_2-albite ($NaAlSi_2O_8$) melt under the nitrogen pressure 1 bar – 3 kbar and the temperature 1250°. Nitrogen solubility with oxygen fugacity (by using a flux Fe-FeO or Ni-NiO) was investigated in magmatic systems of the Earth's mantle [9.33].

9.5.2   Rare Earth Element Alloys

We consider here a few examples of analysis of alloys of rare-earth elements by means of thermal neutrons and gammas from the IPP microtron [9.34]. A bremsstrahlung with maximum energy 29 MeV and thermal neutrons with a flux density $(1 \div 3) \times 10^9$ n/(s·cm²) were used for irradiation of samples. The bremsstrahlung was converted to a fast neutron flux with the aid of a lead converter. A water moderator was used to slow down neutrons to a thermal energy level. Induced activity of the specimens was registered by a semiconductor Ge(Li) spectrometer characterized by a detector volume of 21 cm³ and a spectrometer resolution of 3 keV along 662 keV γ-line of ^{137}Cs.

The sensitivity of determination was first established experimentally by irradiating quantities of elements which are close to the detecting limits (Tables 9.7 and 9.8). The detection limits was assumed to be the value of mass proportional to a useful signal which may be measured by a spectrometer with a relative standard deviation of 0.3.

Table 9.7. The sensitivity of determination of elements by irradiation by bremsstrahlung of 29 MeV electrons from the microtron with average current 10 μA and irradiation time 10 min.

Element	Sensitivity, g
V,Se	$n \times 10^{-4}$
Al,Sc,Cr,Zn,Ni,Rh,Hf,W,Pt	$n \times 10^{-5}$
Co,Ge,As,Nb,In,Te,Gd,Sm,Re,Ir,Pb	$n \times 10^{-6}$
Cu,Ga,Zr,Mo,Sn,Ta,Au	$n \times 10^{-7}$

Table 9.8. The sensitivity of determination of elements at the irradiation by thermal neutrons of the microtron with a flux $\sim 10^9$ n/(s·cm^2), irradiation time 10 min.

Element	Sensitivity, g
Ge,As,Nb,Pr,Gd	$n \times 10^{-4}$
Nd,Al,Ce,Ti,Se,Rb,Sr,Mo,W,Rh,Pd,Sb, Te,Ba,La,Yb,Hf,Hg	$n \times 10^{-5}$
Sc,Cu,Ga,Br,Sm,Ho,Er,Lu,Re,Ir,Th	$n \times 10^{-6}$
V,Mn,Co,I,Au,U	$n \times 10^{-7}$
In,Eu,Dy	$n \times 10^{-8}$

9.5.2.1   Determination of Germanium in Thermal Neutron-Irradiated specimens made of Nb-Ge alloys

Specimens weighing some 30 to 50 mg were wrapped in aluminium foil and irradiated for 20 min in a 1×10^9 n/(s·cm^2) neutron flux. Germanium references (standards) were irradiated at the same time. The weight of the Ge specimens was 25 to 30 mg. The germanium content was determined according to the 0.265 MeV γ-line of ^{75}Ge, and the results are listed in Table 9.9. To illustrate the accuracy of the applied method, the results of chemical analytical analysis and activation analysis performed in a steady reactor are also given.

9.5.2.2   Determination of Germanium and Niobium in Nb-Ge Alloy on Irradiating the Specimens with Bremsstrahlung

Germanium was determined using ^{75}Ge formed as a result of the ^{76}Ge$(\gamma, n)^{75}$Ge reaction, while niobium was the ^{92}Nb resultant from the ^{93}Nb$(\gamma, n)^{92}$Nb reaction. The PAA revealed the presence

Table 9.9. The results of determination of Ge (in %) in Nb-Ge alloys (the confidence intervals are presented for the confidence probability 0.95).

Sample	NAA at microtron	NAA at reactor	Chemical method
1	20.1 ± 0.7	19.8 ± 0.6	21.4 ± 1.0
2	22.8 ± 0.7	22.4 ± 0.6	21.2 ± 1.0
3	27.0 ± 0.9	26.7 ± 0.8	27.4 ± 1.0
4	30.0 ± 0.9	30.3 ± 0.8	31.6 ± 1.3

in specimen 1 (Table 9.9) of $20.4 \pm 0.6\%$ germanium and $80.0 \pm 2.9\%$ niobium, and in specimen 4 of $30.5 \pm 0.9\%$ germanium and $68.9 \pm 2.5\%$ niobium.

The procedure of determining the alloy composition on irradiating the specimens by γ-quanta exhibits a lower error due to a grater number of pulses selected in the analytical peak of ^{75}Ge, and it makes possible to control the accuracy of the analysis of each specimen (in our case, the total of niobium and germanium concentrations should be equal to 100%).

A study of gamma-spectra of the radioactive isotopes resultant from (γ, n) reactions shows that ^{92}Nb and ^{75}Ge analytical lines which were selected cannot be superposed uncontrollably by γ-lines of radioactive isotopes of other elements.

9.5.2.3 Determination of Zirconium in Zr-Nb Alloys

The determination of zirconium in niobium in neutron fluxes supplied by reactors is virtually impossible due to the low activation cross section of zirconium and, mainly, due to interference from tantalum which is normally present in the specimens. Even as low as $(2 \div 3) \times 10^{-3}\%$ the tantalum content inhibits any determination of zirconium, if its concentration in the specimen is below 2 or 3%. Zirconium isotopes have an activation resonance, bur cadmium-plating of the specimens against tantalum-induced interferences cannot produce any significant sensitivity gain, whereas the analysis becomes much more complicated. On irradiating the specimens by γ-rays the sensitivity of zirconium determination (using ^{89m}Zr) is $5 \times 10^{-4}\%$ for niobium, and, as it was shown by some experiments, tantalum concentrations as high as several percents could not hinder determination of such amount of zirconium. Figure 9.12 illustrates this situation. The determination of fractions of a percent and whole percents of zirconium content in an alloy may be performed with high precision, since the number of pulses picked-up in the ^{89m}Zr peak is equal to $n \times 10^3$ to $n \times 10^4$, while the background-to-useful signal ratio is about 30.

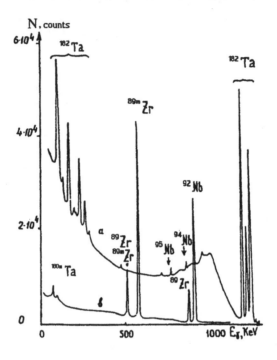

Figure 9.12. Gamma-spectrum of Nb-Zr alloy as a result of the reactions of neutrons (a) and bremsstrahlung (b) irradiation (from [9.34]).

In a study of zirconium distribution (Zr content of about 0.8 to 2%) in ingots and rolled products manufactured from Nb-Zr alloys, it is very convenient to use ^{92}Nb as the monitor (internal standard).

9.5.2.4 Determination of Mo and Zr in a multi-component alloy

Alloy specimens weighing some 40 to 50 mg and containing Mo, Zr and W as alloying admixtures were irradiated by bremsstrahlung for 15 to 20 min. Some 2 to 3 min after irradiation, zirconium was determined as per ^{89m}Zr ($E_\gamma = 589$ keV). In neutron-irradiated specimens molybdenum was determined as per ^{101}Tc ($E_\gamma = 304$ keV), while in gamma quanta irradiated specimens the determination was done as per ^{99}Mo ($E_\gamma = 140$ keV). The results obtained show the same accuracy as above on the level of some percents, and good agreement with the chemical analytical method.

Hence, the PAA of Ge and Zr content in REE alloys permits not only to determine the main components and alloying additives in the specimens, but also the impurities contained in pure substances at concentration levels of 10^{-3} to $10^{-5}\%$. For elements with short-lived radioactive isotopes a relatively high sensitivity may be reached even when irradiating specimens with a low neutron flux (as compared to that supplied by nuclear reactor) from an electron accelerator like a microtron. This fact is accounted for by the possibility of attaining in practice during irradiation values which are close to saturation, or at least comparable with the saturation activity value.

On irradiating the specimens with high-energy bremsstrahlung, problems may be solved which are traditionally considered to be difficult for instrumental neutron-activation analysis. For instance, after irradiation for 6 min, zirconium was determined in hafnium and niobium at sensitivities of $8 \times 10^{-5}\%$ and $5 \times 10^{-4}\%$ respectively, while niobium was found in tantalum at a sensitivity of $5 \times 10^{-4}\%$ (after irradiation for 30 min).

9.6 Activation Analysis of Geological Objects

Rocks and ores are the most difficult objects to analyse due to complexity of their elemental content, inhomogeneous distribution of some elements in the bulk, and the large range of concentrations of elements to be analysed ($n \times 10^1 \div n \times 10^{-8}$ mass.%). The average crustal abundance of ore-forming elements and energy thresholds of the corresponding photonuclear activation reactions are given in Table 9.10.

In addition to the data given it is necessary to underline that the main rock-forming elements have threshold energies of photoactivation that are higher than the corresponding energies for ore elements. For example, carbon has the threshold energy of (γ, n)-reaction 18.7 MeV, oxygen – 15.5 MeV, Na – 12.5 MeV, Mg – 16.6 MeV, Al – 13 MeV, Si – 17.2 MeV, K – 17.2 MeV, Ca – 10.1 MeV, Ti – 13.2 MeV, Fe – 13.8 MeV. So, the data given in the table readily illustrate the main advantages of photonuclear methods in the analysis of rocks and ores as it is possible to avoid radioactivation of matrices in the determination of ore-forming elements by varying the energy of bremsstrahlung.

Table 9.10. Natural abundance of ore-forming elements

Element	Atomic number	Average content in the crust, %	Commercial abundance, %
Li	3	$3 \cdot 10^{-3}$	10^{0}
Be	4	$3.8 \cdot 10^{-4}$	10^{-2}
Sc	21	$1 \cdot 10^{-3}$	10^{-2}
V	23	$9 \cdot 10^{-3}$	10^{-2}
Cr	24	$8.3 \cdot 10^{-3}$	$3 \cdot 10^{1}$
Mn	25	$1 \cdot 10^{-1}$	$3 \cdot 10^{1}$
Co	27	$2 \cdot 10^{-3}$	10^{-2}
Ni	28	$5.8 \cdot 10^{-3}$	10^{0}
Cu	29	$4.7 \cdot 10^{-3}$	10^{0}
Zn	30	$8.3 \cdot 10^{-3}$	10^{0}
Ga	31	$1.9 \cdot 10^{-3}$	10^{-2}
Ge	32	$1.4 \cdot 10^{-4}$	10^{-2}
As	33	$1.7 \cdot 10^{-1}$	10^{1}
Se	34	$5 \cdot 10^{-6}$	10^{-3}
Br	35	$2.1 \cdot 10^{-4}$	10^{-2}
Sr	38	$3.4 \cdot 10^{-2}$	10^{0}
Y	39	$2.9 \cdot 10^{-3}$	10^{-1}
Zr	40	$1.7 \cdot 10^{-2}$	10^{-1}
Nb	41	$2 \cdot 10^{-3}$	10^{-2}
Mo	42	$1.1 \cdot 10^{-3}$	10^{-2}
Ag	47	$7 \cdot 10^{-6}$	10^{-3}
Cd	48	$1.3 \cdot 10^{-5}$	10^{-2}
In	49	$2.5 \cdot 10^{-5}$	10^{-3}
Sn	50	$2.5 \cdot 10^{-4}$	10^{-2}
Sb	51	$5 \cdot 10^{-5}$	10^{0}
I	53	$4 \cdot 10^{-5}$	—
Cs	55	$3.7 \cdot 10^{-4}$	10^{-2}
Ba	56	$6.5 \cdot 10^{-2}$	10^{1}
REE	57-71	$10^{-3} - 10^{-4}$	10^{-1}
Hf	72	$1 \cdot 10^{-4}$	10^{-3}
Ta	73	$2.5 \cdot 10^{-4}$	10^{-2}
W	74	$1.3 \cdot 10^{-4}$	10^{-2}
Re	75	$7 \cdot 10^{-8}$	10^{-4}
Ir	77	$1 \cdot 10^{-6}$	—
Pt	78	$2 \cdot 10^{-5}$	10^{-4}
Au	79	$4.3 \cdot 10^{-7}$	10^{-4}
Hg	80	$8.3 \cdot 10^{-6}$	10^{-1}
Pb	82	$1.6 \cdot 10^{-3}$	10^{0}
Bi	83	$9 \cdot 10^{-7}$	10^{-2}
Th	90	$8 \cdot 10^{-4}$	10^{-2}
U	92	$2.5 \cdot 10^{-4}$	10^{-2}

9.6.1 Nondestructive Determination of Deuterium and Beryllium

Previously we have mainly considered activation with a nuclear reaction of the type (γ, n). The thresholds of these reactions are in general higher than 8 MeV. There are only two exceptions, that of beryllium and deuterium which have threshold energies 1.67 and 2.23 MeV, correspondingly. Thus by irradiating photons of only a few MeV, one can produce neutrons arising from these elements. This gives rise to a possible analytical application relating to the determination of beryllium and deuterium by counting the photoneutrons emitted during irradiation.

In particular, Mazyukevich and Shkoda-Ul'yanov [9.35] have discussed the possibility of determining hydrogen in metals, utilizing deuteron photodisintegration They estimate that with a 50-μA electron beam of 4.5 MeV in energy, metals containing as little as 10^{-4}% hydrogen can be analysed. It is necessary that the beryllium level in the sample be two or three orders of magnitude lower than that of hydrogen. On the other hand, deuteron photodisintegration can be applied for investigating isotopic fractionation of hydrogen in natural water [9.36]. Overman *et al.* [9.37] have used the reaction to monitor heavy-water production and have used deuterium oxide in tracing soil moisture. Guinn and Lukens [9.38] have demonstrated that bremsstrahlung from a 1 mA beam of 3 MeV electrons provides the sensitivity necessary for deuterium determination to 1/500 of its concentration in natural water.

The isotope sources were predominantly used for determination of beryllium, and, in particular, as a portable instrument for field work. The advancement of accelerator techniques made it possible to use a modest-sized electron accelerator in field analysis of beryllium. In particular, a 3 cm-microtron with the electron energy 4 MeV was specially designed and built for geophysical researches [9.39]. The vacuum chamber of the microtron is 205 mm in diameter, and the authors used 200 kW tunable magnetron as a rf source. The accelerator operated in the second type of acceleration with the parameter $\Omega \sim 0.6$, $\epsilon \sim 1.1$ having been proposed and realized earlier by Rodionov and Stepanchuk [10.6] The pulse current on the 13th orbit was ~ 1 mA (mean current 1 μA), pulse length was 0.65 μs, repetition rate 100–800 Hz.

To fully use the analytical possibilities of the microtron the authors also paid careful attention to accurate design of the detecting system. The main problem in a low-background neutron detecting system is usually due to detector sensitivity to gamma-rays originating in the irradiation facility, and due to the background from cosmic rays (muons, neutrons, and electron-photon avalanches). In the experiments discussed the authors used a scintillation detecting system, and to make the detector insensitive to bremsstrahlung they locked in the photomultiplier by feeding a pulse of negative voltage for $10\,\mu s$ in synchronism with the electron beam.

Specimens containing the beryllium compound $Be(NO_3)_2 \cdot 4H_2O$ of $800\,cm^3$ in volume were irradiated by bremsstrahlung from tantalum target placed inside the microtron, the distance from target to sample was $300\,mm$. To test the background of the analysing system the compound H_2NCONH_2 was used instead of a beryllium containing sample. The geometry of the experiment and the detecting system are given in Fig. 9.13.

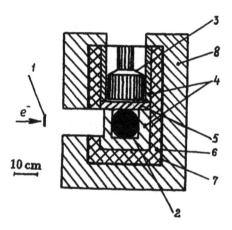

Figure 9.13. Detecting system for registration of photoneutrons emitted by a beryllium containing sample: 1 — bremsstrahlung target, 2 — probe, 3 — detector, 4 — paraffin neutron reflector, 5 — plexiglass moderator, 6 — lead shielding, 7 — tetraboronacid sodium, 8 — plexiglass shielding.

Under photon irradiation of the probe, photoneutrons with energies of some MeV are originated due to the (γ, n) reaction. A photomultiplier with a neutron-to-photon scintillation converter in the front window is used as a neutron detector. To increase the count rate the detector and the sample are surrounded by paraffin reflector $50\,mm$ thick. The registration of neutrons begins with a time delay ($\sim 200\mu s$) relative to the electron beam pulse. The plexiglass moderator $\sim 4\,cm$ in thickness is placed between the sample and the detector. The $5\,cm$ lead screen serves for shielding the scintillator and the photomultiplier from bremsstrahlung, and protection from the neutron background is provided by a $4\,mm$ layer of tetraboronacid sodium and by $100\,mm$ plexiglass blocks.

The experiments showed that at 30 min measurements the detection limit of the method discussed equals $(1-2) \times 10^{-4}$ mass %, corresponding to modern geological demands.

9.6.2 Isomer Production Under Bremsstrahlung Irradiation

The existence of nuclear isomers which can be produced by the reactions of the type (γ, γ') leads to very interesting analytical possibilities. The great advantage of this method arises from the fact that using rather low energy bremsstrahlung (~ 10 MeV) only a few elements are activated, and all these isomers are pure gamma emitters. Thus the corresponding element can be determined in a nondestructive manner. Energy dependence of isomer yields experimentally measured at a linac [9.40] are shown in Fig. 9.14 for elements suitable for use in express activation analysis. Their characteristics are listed in Table 9.11.

Table 9.11. Detection limits of elements in gamma-activation analysis by measuring γ-radiation of isomeric nuclei produced

Element	Half life, s	Energy of γ-line, keV	Detection limit, %
Au	7.2	279	4×10^{-5}
W	5.3	46.53	9×10^{-4}
Ag	44.3	87.93	2×10^{-4}
Se	17.5	161	3×10^{-4}
Br	4.8	210	2×10^{-4}
Er	2.3	208	5×10^{-4}
Hf	18.6	217	4×10^{-5}
Ba	156	662	5×10^{-4}
Y	16	910	3×10^{-4}

The detection limits are measured at electron energy 8 MeV, the average current 700 μA, and the mass of the sample 500 g.

Nuclear isomers are widely used in analytical problems, and we consider below two directions for applying this technique — the analysis of gold and silver content in goldbearing ores and express analysis of a composition of mineral sources.

9.6.3 Express Activation Analysis of Gold and Silver

The abundance of gold and silver in nature is severely limited. The mean clarke content of these elements are 5×10^{-7} and 1×10^{-5} mass.% correspondingly. In addition, these elements,

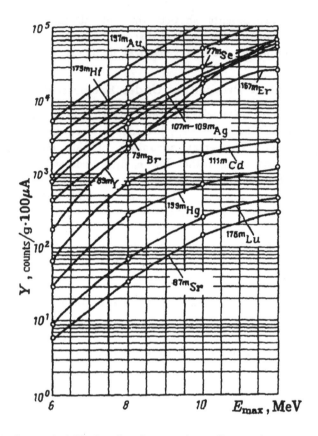

Figure 9.14. Integral yield of nuclear isomers depending on electron energy; the data are normalized to 1 g of element and average electron current 100 μA (from [9.40]).

and especially gold, are nonuniformly distributed in minerals, and this fact leads to the need to analyse probes of large mass (tens or even thousands of grams). Besides, large-scale analyses, needed for geology and industry, require high-accuracy and express analytical methods for determining noble elements. Activation methods make it possible to realize a technique fully satisfying the given requirements. In particular, PAA is one of the most developed technique for solving this task.

The express variant of PAA is based on the use of short-lived isomers of gold ^{197m}Au and ^{107m}Ag, ^{109m}Ag, which are formed as a result of photoexcitation (reaction of inelastic scattering of gamma quanta) of corresponding stable isotopes. The half-life ($T_{1/2}$) of ^{197m}Au equals 7.3 s, and the energy of emitting γ-

rays equal 70, 130 and 277 keV. The most intensive are 10 and 277 keV lines, which are used for gold identification. The nuclear characteristics of silver isomers (^{107m}Ag — 44 s, 94 keV, ^{109m}Ag — 39 s, 88 keV) are close in value and are not usually separated in the processing of experimental information.

The experiments performed [9.41] show that it is possible to achieve the sensitivity of express PAA for gold 1 g/t, and 10 g/t for silver that fully satisfies industrial demands, and the technique was promoted in several geological regions of the USSR. As an example, Fig. 9.15 presents the gamma spectrum of induced activity in a quartz-carbonate ore with the gold content 14 g/t and silver content 76 g/t measured with the NaI(Tl)-detector.

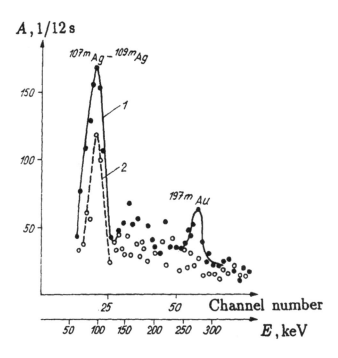

Figure 9.15. Gamma-spectrum of a gold-bearing ore with the gold content 14 g/t and silver content 76 g/t: 1(●) — cooling time equals 4 s, 2(o) — 32 s; detector — NaI(Tl) 70×70 mm, electron energy — 8.5 MeV, average current 50 μA.

The photopeak 277 keV is due to the activity of the gold isomer ^{197m}Au, which is used for the quantitative calculation of the gold content, whereas in the energy region 88–94 keV the activity of isomers ^{107m}Ag and ^{109m}Ag as well as the 70 keV gold photopeak are presented. That is why for silver content determination second measurement of the induced activity is necessary, and cooling the sample for 32 seconds is enough for gold activity to decay (curve 2 in Fig. 9.15).

It is necessary to note that the existence in the goldbearing ores of other elements is of great importance. For example, Ba and Hf are paragenetically connected in some fields with Au and are a geochemical criterium at ore searches, Se and Ba together with other sulphides are often present in goldbearing ores and are the products of secondary processing of a source.

To determine the optimal conditions of gold determination in ores the experiments [9.42] with the ores from different fields of the USSR are analysed at different electron beam energies by the IPP microtron under the following conditions: irradiation time was equal to 20 s, cooling time 5 s, measurement time 20 s. As a gamma quanta detector a Ge(Li) detector was used.

In accordance with their mineralogical composition the samples were divided into two groups: 'simple' in which silica and alumosilicates with insufficient content of sulphides and oxides of different metals are the main part of the matrix, and 'complex' where metal sulphides and oxides are at a considerable (1-10%) level. The gamma-spectra of induced γ-activity in typical simple and complex samples are presented in Fig. 9.16 and 9.17. Irradiations were made with bremsstrahlung at three endpoint energies – 8, 10, and 15 MeV.

Photopeaks of Ag (90 keV), Se (162 keV), Hf (217 keV) and Au (279 keV) are distinctly seen in the figures presented. It is also seen from Fig. 9.16 that in a complex probe, even at the energy 8 MeV, the photopeaks from Au and Ag are brought by the Compton pedestal from the Ba γ-line that essentially decreases the sensitivity of the analysis; and in a complex probe an analogous pedestal originated at higher energies from the annihilation peak.

9.6.4 PAA of Multicomponent Ores

Activation analysis is usually applied for analysing a small amount of impurity or trace elements in different objects. Now we consider the problem of determination of great amounts of elements in ore but with high accuracy and expressness. In particular, this problem appears in evaluation of the quality of multicomponent ores. We consider the results of applying express PAA to the analysis of apatite nepheline ores [9.43], whose elemental composition is rather simple. The major useful components of these ores are phosphorous, aluminium and titanium.

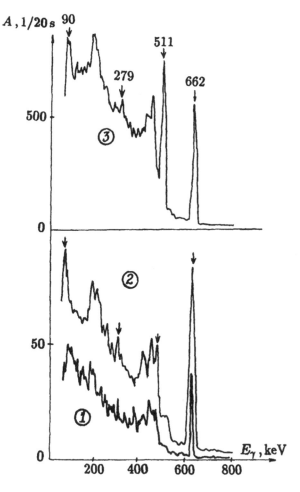

Figure 9.16. Gamma-spectrum of the complex goldbearing sample analysed at the bremsstrahlung end point energies 8 (*1*), 10 (*2*) and 15 (*3*) MeV.

The following main short-lived nuclides are formed by bremsstrahlung in the useful components as well as in cerium and barium:

$$^{31}\text{P}(\gamma, n)^{30}\text{P} \qquad T = 153\,\text{s}, \qquad E_\gamma = 0.511\,\text{MeV},$$
$$^{27}\text{Al}(\gamma, n)^{26\text{m}}\text{Al} \qquad T = 6.7\,\text{s}, \qquad E_\gamma = 0.511\,\text{MeV},$$
$$^{47}\text{Ti}(\gamma, p)^{46\text{m}}\text{Sc} \qquad T = 19.5\,\text{s}, \qquad E_\gamma = 0.140\,\text{MeV},$$
$$^{140}\text{Ce}(\gamma, n)^{139\text{m}}\text{Ce} \qquad T = 55\,\text{s}, \qquad E_\gamma = 0.740\,\text{MeV},$$
$$^{138}\text{Ba}(\gamma, n)^{137\text{m}}\text{Ba} \qquad T = 150\,\text{s}, \qquad E_\gamma = 0.662\,\text{MeV}.$$

Along with it, such oreforming elements as oxygen, silicon,

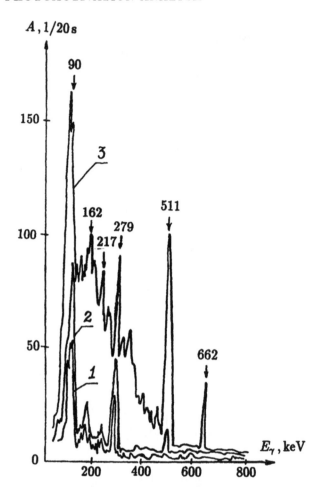

Figure 9.17. Gamma-spectrum of the simple goldbearing sample analysed at the bremsstrahlung end point energies 8 (*1*), 10 (*2*) and 15 (*3*) MeV.

calcium and others are not activated at all, or slightly activated at the energies of activating photons less than 15 MeV. The experiments, performed at the IPP microtron, were characterized by the following conditions: the irradiation time was equal to 20 s at electron energy 15 MeV and average current 10 μA, after 5 s (the time needed for transporting the sample from the accelerator to the detecting system) the induced γ-activity was measured by a Ge(Li)-detector in two successive intervals of 20 s each; the mass of the sample was equal to approximately 100 g.

Figure 9.18 shows the spectra of three typical probes of apatite nepheline ores.

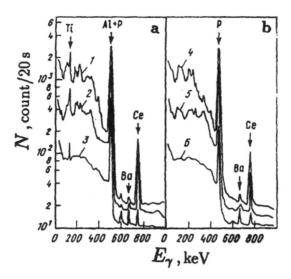

Figure 9.18. Gamma-spectrum of the apatite nepheline ores irradiated by bremsstrahlung with the end point 15 MeV: a — cooling time 5 s; b — cooling time 30 s.

The content of the specimens was:

$1,4$ – P_2O_5 – 16.1%, TiO_2 – 16.6%, Ce_2O_3 – 0.26%;
$2,5$ – P_2O_5 – 11.4%, TiO_2 – 6.3%, Ce_2O_3 – 0.17%;
$3,6$ – P_2O_5 – 1.1%, TiO_2 – 2.1%, Ce_2O_3 – 0.04%.

Based on the experimental results the calculation showed that the sensitivity of the simultaneous determination of P, Al, Ti, Ba, and Ce is 0.1, 0.1, 0.1, 0.1, and 0.001 mass % respectively. The statistical error of the analysis in the ore with 0.1% Ce, 3% Ti, and 5% P was 10, 10, and 3% respectively. These values wholly satisfy to modern demands of industry, and can easily be improved by increasing the detection efficiency as may be required.

9.6.5   PAA of Polymetal Ores in a Mixed Gamma-Neutron Field

Express activation analysis of a mineral source performed in a mixed gamma-neutron field (see Chapter 7) essentially increases the amount of elements which can be determined. For instance, sodium, manganese, vanadium are slightly activated by γ-quanta but well activated by thermal neutrons, and in contrast lead and zirconium are difficult objects for NAA but are well activated by γ-quanta. Such a mixed $(\gamma - n)$-field was realized at the microtron of IPP (Moscow) to study the possibilities of express analysis of geological specimens [9.44]. Figure 9.19 shows a schematic diagram of the experiments.

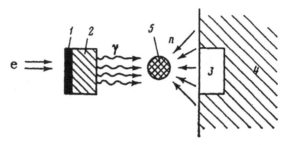

Figure 9.19. The diagram of obtaining a $(\gamma - n)$-field at the IPP microtron: 1 — bremsstrahlung target, 2 — aluminium absorber, 3 — lead neutron converter, 4 — hydrogen containing moderator (polyethylene), 5 — irradiated sample.

Gamma quanta originated in a tungsten target by 15 MeV electrons pass through a sample and then enter a thick lead converter $10 \times 10 \times 5$ cm in size. The converter is placed in a large polyethylene moderator, and thus in the sample irradiation position we have bremsstrahlung, fast, resonance and thermal neutrons simultaneously. The thermal neutron flux density in such a geometry equals $\sim 10^6$ neutr/(cm²·s·µA). In addition, in the irradiation zone there is one order of magnitude higher flux of resonance neutrons. These fluxes are enough for determination of V, Dy, In, and Mn with sensitivities which are comparable with those for gamma quanta obtained. Figure 9.20 illustrates the possibilities of the method by the example of a polymetal ore.

9.6.6 Gold Content Determination in Geological Samples with the Use of Resonance Neutrons

Gold activation can be performed in several ways. The most widespread techniques are the following:

(i) the discussed above excitation of the isomeric state of gold ^{197m}Au ($T_{1/2} = 7.2$ s) whose transition to the ground state leads to a γ-quantum emission with the energy 279 keV;

(ii) use of the (γ, n)-reaction producing long-lived gold activity;

(iii) use of the reaction of neutron radiative capture ^{197}Au$(n, \gamma)^{198}$Au with a formation of β-active nuclide ^{198}Au ($T_{1/2} = 42.7$ d) whose decay is accompanied by emission of γ-quanta with the energy 411.9 keV;

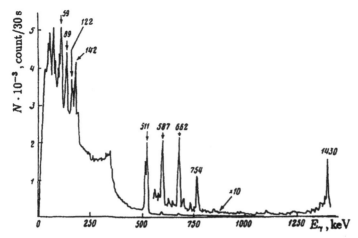

Figure 9.20. Gamma spectrum of polymetal ore obtained by express PAA in a mixed $(\gamma - n)$-field at electron energy 15 MeV and registered by a Ge(Li)-detector; mass of the sample is 100 g, irradiation time — 30 s, cooling time — 5 s, measurement time — 30 s. The intense photopeaks correspond to ^{185m}W (59 keV), ^{178}Hf (89 keV), ^{152m}Eu (122 keV), ^{46m}Sc (142 keV), ^{89m}Zr (587 keV), ^{137m}Ba (662 keV), ^{139m}Ce (754 keV), ^{52}V (1430 keV).

Using energetic bremsstrahlung one can excite the long-lived gold isotope ^{196m}Au with $T_{1/2}$=6.2 d and E_γ=333.36 keV by (γ, n)-reaction [9.41]. Figure 9.21 shows the result obtained by 14 MeV-bremsstrahlung irradiation of the sample during one hour with average current 25 μA.

The gamma spectrum was measured with a Ge(Li) detector. In the spectrum obtained there is information not only about the gold content but also photopeaks associated with Pb, As, Ba, Sb, Cu, Mn, and Fe, giving additional useful information about the probe examined. The sensitivity of the analysis by means of long-lived gold radionuclide is better than 0.1 g/t, giving wide possibilities for geological investigations.

The highest sensitivity of activation analysis can be reached in the case when activation of the matrix is minimal. In the case of thermal neutron activation analysis the use of resonance neutrons gives the best result. The thermal neutron cross section for gold has well pronounced resonance at neutron energy 5.12 eV with a maximum value $\sim$ 1550 barn (see Fig. 9.22), and this feature enables the successful use of resonance neutrons for efficient gold determination in rocks and ores. In particular, this method was applied by Belov *et al.* [9.45] for analysing geological samples by means of a graphite cube (see Ch.7). To decrease the level of

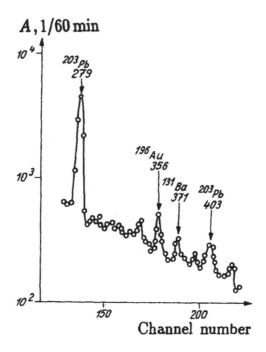

Figure 9.21. Gamma-spectrum of a polymetal ore with the gold content 0.4 g/t after 30 h cooling; $E_{max} = 14$ Mev, $\bar{I} = 25 \mu A$.

matrix activation the samples were screened by cadmium sheets that significantly increased selectivity of the analysis.

Samples with monitors (sheets of paper impregnated by a solution with known gold content) were irradiated for 10 hours, and measurements of their activity were performed after 90-100 hours. As a result of cooling the samples the activity of such nuclides as ^{24}Na and ^{56}Mn practically disappeared, and the activity of the nuclide ^{76}As is also considerably decreased. The main interferences occur from arsenic and stibium which have γ-lines 559 and 564 keV respectively. The edge of the Compton distribution from these lines is superimposed on the analytical line 412 keV from gold-198, and therefore the sensitivity and measurement error are determined by the content of these elements in samples to be analysed.

The maximum sensitivity of the method was 0.1 g/t ($10^{-5}\%$). Several thousand samples were analysed by the authors, the results were compared with chemical analysis data, and to illustrate the reproducibility and accuracy of the neutron activation method

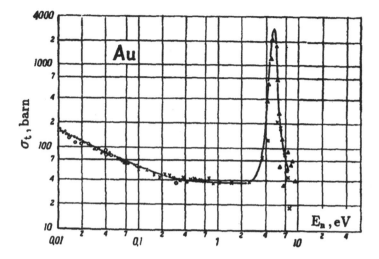

Figure 9.22. Neutron cross section for gold.

results in relation with chemical data is shown in Fig. 9.23.

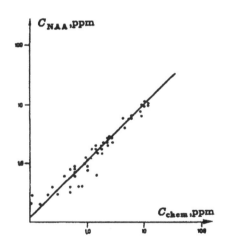

Figure 9.23. The correlation between chemical and neutron-activation methods of analysis.

As it was mentioned above, the sensitivity and accuracy of the method are determined by the background in the region of the gold analytical line 412 keV. Thus, decreasing the background is the main way to increase the sensitivity. As the most radical way, it is possible to use preliminary chemical separation of interfering elements of the matrix (As and Sb). The second way to decrease the background is screening of samples by absorbers made with mixed arsenic and stibium, but this way, unfortunately, gives a rather small effect (background only $1.5-2$ times smaller). Nevertheless, estimation shows that it is possible to increase the sensitivity of the method up to $2 \cdot 10^{-8}$ g/g by using a more efficient detection system.

To enhance the detection efficiency the induced activity of gold (the energy of emitted γ-quanta equals $E_\gamma = 411.8$ keV) a twinned Ge(Li) gamma-ray spectrometer was used, consisting of two Ge(Li) detectors, each $50\,\mathrm{cm}^3$ in volume. This provided a detection efficiency equal to 13%. This technique provides a threshold of gold determination of about 0.03 ppm for an irradiation time of 10 hours and for 10 minutes measuring time for each sample. These characteristics of the method fully satisfy the requirements imposed by most practical problems. As a large number of samples can be irradiated simultaneously the productivity is rather high, reaching 250 samples per day.

9.6.7 Instrumental Determination of the Uranium and Thorium Content in Natural Objects

In NAA analysis of uranium and thorium content is based on the activation of the probe by epithermal neutrons from a reactor or an accelerator with subsequent measurement of γ-activity of ^{239}U ($E_\gamma = 44$ and $74.7\,\mathrm{keV}$) or ^{239}Np ($E_\gamma = 106$, 228 and $278\,\mathrm{keV}$) for uranium determination, and by means of ^{233}Pa activity with $E_\gamma = 97$, 312, 340, 375, 397, and $416\,\mathrm{keV}$ for thorium determination. The detecting limit reaches $1-0.1\,\mathrm{g/t}$ but depends on the composition of the samples analysed [9.46]. For example, the sensitivity is significantly decreased in the case of uranium determination in iron-manganese concretions in which the manganese content can reach 42%.

Another way is to use photon activation [9.47, 9.48]. In PAA of uranium content it is worthwhile to use the reaction

$$^{238}\mathrm{U}(\gamma, n)^{237}\mathrm{U}\ \beta^- \longrightarrow\ ^{237}\mathrm{Np}.$$

In the reaction the following photopeaks are convenient to use for uranium determination: 59.54 and 208 keV, or $E_{\mathrm{K}_{\alpha 2}\mathrm{Np}} = 97.08\,\mathrm{keV}$, $E_{\mathrm{K}_{\alpha 1}\mathrm{Np}} = 101.7\,\mathrm{keV}$.

Figure 9.24 shows the spectrum of the product, irradiated by bremsstrahlung, of processing of iron-manganese concretions with $0.6\,\mathrm{g/t}$ uranium content. The analytical lines of uranium 59.54,

97 and 101 keV are not superimposed by other activity, and the sensitivity equals 0.1 g/t. For comparison note that at neutron activation the detecting limit is only of the order of 1 g/t.

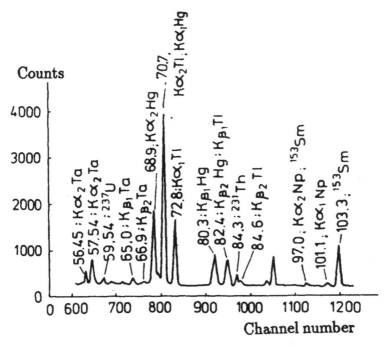

Figure 9.24. Gamma-spectrum of the product of concretion processing with the uranium content 0.6 g/t (from [9.49]).

In the case of thorium determination the corresponding reaction is

$$^{232}\text{Th}(\gamma, n)^{231}\text{Th} \xrightarrow{\beta^-} {}^{231}\text{Pa}.$$

The most convenient analytical line is 84.17 keV. Figure 9.25 illustrates the fragments of the spectra in the region of the thorium analytical line that were obtained after bremsstrahlung irradiation of different samples.

9.6.8 Photofission in Analysis of the Quality of Lapparite Ores

It is necessary to underline that in the processes discussed above the sensitivity of gold and silver determination depends to a large degree on fissionable elements existing in the ore.

On the other hand, an abundance of fissionable elements can be used in some cases for analysis of the quality of examined rocks

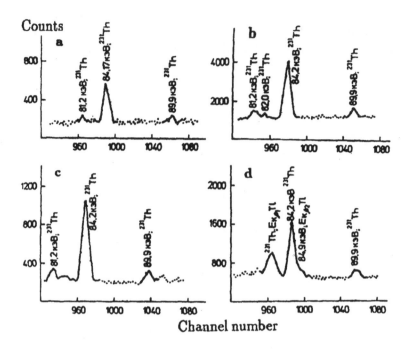

Figure 9.25. Gamma-spectrum of the samples irradiated by a bremsstrahlung from the Dubna microtron: a — geological sample ($C_{Th}=0.8\,g/t$), b — iron laterite ore ($C_{Th}=15.2\,g/t$), c — soil ($C_{Th}=8.8\,g/t$), d — iron-manganese concretions ($C_{Th}=15.1\,g/t$) (from [9.49]).

and minerals. For instance, thorium always exists in lapparite ores and one can see in them characteristic natural gamma radioactivity — a 239 keV gamma line, and the well known method of estimating the quality of a lapparite ore is based on measurement of the level of this radioactivity. The sensitivity and accuracy of the thorium content determination can be significantly improved by measuring photoinduced activity. The photofission cross section of thorium at $E_\gamma \simeq 15\,MeV$ is rather high ($\sim 50\,barn$) and in consequence a lot of radioactive fission fragments are produced even under rather low energy bremsstrahlung irradiation. Besides, most fission fragments have short half-lives and this feature of the fission process is another advantage of applying PAA in the problem discussed.

The following experiment performed at the IPP microtron illustrates these statements. A lapparite ore sample was irradiated by 10 MeV bremsstrahlung for 30 s and then the induced activity was measured by a Ge(Li) detector for 30 s — see Fig. 9.26.

Two essential features of the gamma spectra presented in Fig. 9.26 are clearly seen — the high level of short-lived gamma activity of fission fragments in the energy region 50-500 keV and the existence of a characteristic thorium 239 keV gamma-line in the 60 s-cooling spectrum on a rather small pedestal of long-lived fission fragments. So, in this particular case PAA provides a very promising express method of evaluating the quality of lapparite ores in solving any kind of geological problems.

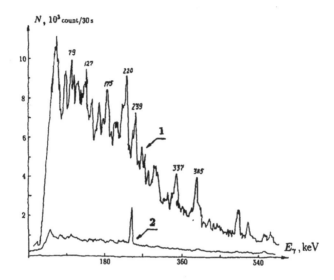

Figure 9.26. Gamma spectra of a lapparite sample irradiated by 10 MeV bremsstrahlung: 1 — cooling time equals 5 s, 2 — cooling time equals 60 s.

Chapter 10

Microtrons in Medicine

Nuclear science has been and remains a major contributor to various areas of medicine. Various radionuclides are used for diagnosis and treatment of deseases and a lot of medical and biological investigations, nuclear analytical methods are used to determine the content of the different microelements in human organism, human nutrition and environment, labeling allows us to study *in viva* functioning of different organs. Experts hold that for normal functioning a person needs different quantities of about 15 main microelements such as iodine, iron, copper, zinc, cobalt and selenium, and radioactivation methods of analysis, in particular the photoactivation technique, play a major role in these studies.

Certainly, a very essential direction of applying microtrons in medicine is their use for cancer therapy, to which the next chapter is devoted.

10.1 Analysis of Biological Objects

Analysis of biological objects is one promising direction of application of PAA. One of the main disadvantage of NAA is the high activation of biological macroelements (Na, K, Cl) that makes it impossible to use a nondestructive (instrumental) variant of analysis for radionuclides with half life less than 1 day. Photoactivation analysis is free from this limitation. It is also necessary to pay attention to the possibility of performing selective analysis due to the different threshold character of photonuclear reactions. Practically all elements are involved in the human body and animals, and most of them are vitally important. Activation analysis encourages diagnosis at early stages of deseases and control of the efficiency of therapeutic treatments.

279

The feasible detection limits of some elements in human blood with instrumental PAA are listed in Table 10.1 to illustrate these advantages.

Table 10.1. Detection limits of some elements in blood by instrumental photoactivation analysis (from [10.1]).

Element	Photonuclear reaction	Radionuclide parameters		Detection limit		Native concentration
		$T_{1/2}$	E_γ	μg	μg/ml	μg/ml
C	$^{12}C(\gamma,n)^{11}C$	20.3 m	511	0.23	1 – 10	23636
N	$^{14}N(\gamma,n)^{13}N$	10.0 m	511	0.73	1 – 10	25000
O	$^{16}C(\gamma,n)^{15}O$	124 s	511	0.13	1 – 10	745455
F	$^{19}F(\gamma,n)^{18}F$	110 m	511	0.034	10	0.173
Na	$^{23}Na(\gamma,n)^{22}Na$	2.6 y	1275	2400	180	1818
			511	392	-	
Mg	$^{25}Mg(\gamma,p)^{24}Na$	14.9 h	1368	13	12	24
Cl	$^{35}Cl(\gamma,n)^{34m}Cl$	32 m	146	0.38	15	2727
			511	0.40	-	
K	$^{39}K(\gamma,n)^{38m}K$	7.7 m	2160	4.0	-	1600
Ca	$^{44}Ca(\gamma,p)^{43}K$	22.4 h	372	39	45	56
Fe	$^{57}Fe(\gamma,p)^{56}Mn$	2.58 h	847	12	25	455
Ni	$^{58}Ni(\gamma,n)^{57}Ni$	36 h	511	0.45	-	0.029
			1378	1.6	0.8	
Zn	$^{68}Zn(\gamma,p)^{67}Cu$	59 h	185	12	11	6.2
As	$^{75}As(\gamma,n)^{74}As$	17.9 d	596	5.4	8	0.45
Rb	$^{85}Rb(\gamma,n)^{84}Rb$	33 d	881	13	4	2.6
Sr	$^{88}Sr(\gamma,n)^{87}Sr$	2.83 h	388	0.02	0.5	0.033
Zr	$^{90}Zr(\gamma,n)^{89}Zr$	78.4 h	511	2.3	-	2.4
			913	2.1	0.8	
Mo	$^{100}Mo(\gamma,n)^{99}Mo$	67 h	181	16	8.5	0.0043
Cd	$^{116}Cd(\gamma,n)^{115}Cd$	53 h	336	-	6.5	0.0065
Sb	$^{123}Sb(\gamma,n)^{122}Sb$	2.8 d	564	1.0	1.0	0.0044
Cs	$^{133}Cs(\gamma,n)^{132}Cs$	6.5 d	668	0.88	0.5	0.0027
Hg	$^{204}Hg(\gamma,n)^{203}Hg$	46.9 d	279	27	25	0.0047
Tl	$^{203}Tl(\gamma,n)^{202}Tl$	12 d	439	3.3	1.5	0.09
Pb	$^{204}Pb(\gamma,n)^{203}Pb$	52.1 h	279	6.4	13	0.255

The results of studying the sensitivity of PAA for a set of biological objects by 45 MeV linac with average current 5–10 μA were also published in [10.2].

10.1.1 Photoactivation Measurement of Oxygen Isotopic Concentration Ratios $^{18}O/^{16}O$

Many different nuclear methods exist for selective oxygen-18 determination, or even for measurement of the $^{18}O/^{16}O$ isotopic concentration ratio. They are based either on direct observation of nuclear reactions induced by charged particles or on the use of γ, neutron or charged particle activation processes. However, a special sample preparation is usually necessary before irradiation, especially when analyzing biological media (blood, urine, etc.) by means of charged particles. It has been The possibilities offered by γ-quanta as a very simple, fast and nondestructive means of performing such analysis were mentioned by Engelmann and Scherle [10.3] and then realized in [10.4].

The method is based essentially on simultaneous or staggered measurements of the activities induced in the above isotopes by the nuclear reactions $^{16}O(\gamma,n)^{15}O$ and $^{18}O(\gamma,p)^{17}N$ of threshold energies 15.7 and 16.3 MeV respectively. Oxygen-15 is a pure β^{+}-emitter of half-life 2.03 minutes; nitrogen-17 is a delayed neutron emitter of half-life 4.2 seconds. The oxygen-15 positron annihilation γ-rays are easy to detect with a sodium iodide crystal, the nitrogen-17 neutron activity can be recorded with BF_3 or ^3He type counters, and these activities can be measured simultaneously.

The activities of these radionuclides are obviously proportional to the quantities of oxygen-16 and oxygen-18 irradiated, which means that the quotient of the two count rates gives the $^{18}O/^{16}O$ isotopic concentration ratio directly, if calibration is carried out beforehand.

Figure 10.1 gives the activation curve for the nuclear reaction $^{18}O(\gamma,p)^{17}N$. The ordinates express the number of neutrons detected per gram of ^{18}O per second at the end of a 30 s irradiation, plotted against the incident electron beam of mean intensity $100\,\mu A$. These experiments were carried out by activation of water samples $(V = 0.5\,cm^3)$ at natural isotopic concentration. The containers were irradiated about 2 mm behind a 7 mm thick platinum target.

The method was used to determine the pulmonary extravascular water volume in rats. The process is based on the injection of a diffusible tracer (0.15 g of oxygen-18-labelled water; enrichment factor — 15%) at the lung input, followed by blood sampling every second at its output. The biological parameter concerned was calculated from data supplied by the variation with time of the oxygen-18 concentration in these samples.

Figure 10.2 shows a typical curve representing the oxygen-18

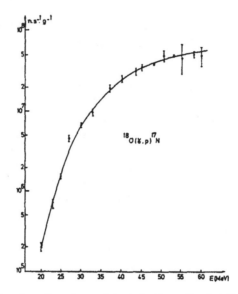

Figure 10.1. Number of neutrons detected after a 30 s irradiation, per gram of oxygen-18, per second, as a function of incident electron beam energy, $I = 100\,\mu$ (from [10.4])

concentration variations in blood samples taken over the first 25 seconds following injection. From its surface area, the required elements are obtained by comparison with a reference vascular indicator injected at the same time.

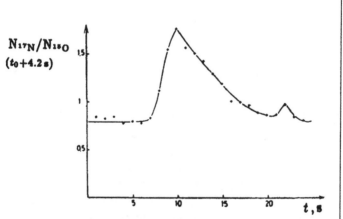

Figure 10.2. Typical curve showing the oxygen-18 content variation with time in blood samples taken at the lung output in rats. The experimental points correspond to successive blood samplings carried out every second after injection of the diffusible tracer upstream from the lung (from [10.4])

This medical research technique could offer interesting possibilities for the study of liquid exchanges in normal and pathological human subjects. The main advantage of the analytical method described is that it can be applied directly to any oxygen-18 labelled molecule without requiring any sample treatment before measurement.

10.2 Radionuclides in Therapy

Medical and clinical applications of radionuclides were among the first, after the discovery of radioactivity. When artificially produced radionuclides became available from irradiation in nuclear reactors or particle accelerators, medical therapy was again among the major fields of interest. Nuclear physics made it possible to produce nuclides with much shorter half-lives than radium and a broad spectrum of decay energies which permitted more flexibility in the irradiation schemes of surgically introduced or implanted radiation sources.

There are more than 170 radionuclides of 85 elements used in medicine, 60% of which are produced in reactors and 40% in cyclotrons. Radioisotope programs are in the working plans of numerous accelerator centers. In particular, at the LAMPF accelerator (Los-Alamos, USA) and at TRIUMF (Vancouver, Canada) for radionuclide production the proton beams passing meson targets, but possessing up to 70% of their initial intensity are used. At such powerful accelerators (they are often named 'meson factories') ^{77}Br(56 h) and ^{123}I(13 h) are produced for radiopharmacology, ^{127}Xe(36 d) for studying lungs, ^{67}Ga(78 h), ^{111}In(2.8 d) and ^{167}Tm(9.3 d) for oncology, ^{82}Sr(25 d) to study dynamics of the circulation of blood, ^{52}Fe(8 h) for the marrow, ^{44}Ti for the skeleton.

As we can see, radionuclides intended for medical use are now produced mostly in reactors or cyclotrons. However, an electron accelerator (EA) can be used for the same purpose. The EA has some advantages. Both neutron deficient and neutron surplus nuclides can be produced in the EA by photonuclear reactions. The bremsstrahlung irradiation heats the target less than charged particle bombardment. This makes target preparation and heat removal easier. Sometimes a higher radionuclidic purity of the preparation can be obtained using a photonuclear reaction rather than a reaction with a charged particle.

McGregor was the first to point out the possibility of using EA for radionuclide production [10.5]. However, practical developments followed later [10.6–10.10].

Experimental yields of some radionuclides produced by the (γ, p) reaction at $E_e = 25$ MeV were measured by Levin *et al.* [10.11] and are given in Table 10.2.

Table 10.2. Experimental yields of some radionuclides produced by the (γ, p) reaction with 25 MeV electron bremsstrahlung (from [10.11]).

Radio-nuclide	Half life, h	Target-nucleus	Isotopic content, %	Yield, μCi/μA·h·g
^{43}K	22.4	^{40}Ca	94.9	4.3 ± 0.5
^{56}Mn	2.6	^{57}Fe	natural	0.3 ± 0.1
^{61}Co	1.6	^{62}Ni	natural	3.5 ± 0.7
^{67}Cu	61.0	^{68}Zn	natural	0.35 ± 0.05
^{63}Ga	4.9	^{74}Ge	natural	0.5 ± 0.1
^{101m}Rh	105.6	^{102}Pd	80	12±3
^{111}In	67.4	^{112}Sn	80	32.–55.5
^{123}I	13.3	^{124}Xe	3.4	10.4±4
^{129}Cs	32.1	^{130}Ba	32.9	32.3 ± 3.5

Below we consider some examples of the methods for medical radionuclide production using electron accelerators.

10.2.1　Carrier-Free Nuclides ^{155}Tb and ^{167}Tm

Isotopes of some rare earth elements were shown to be useful in nuclear medicine [10.12–10.14]. Particularly ^{167}Tm [10.15] and ^{155}Tb were proposed for skeleton imaging. At first ^{155}Tb and ^{167}Tm were produced in the cyclotron via ^{153}Eu$(\alpha, 2n)^{155}$Tb and ^{165}Ho$(\alpha, 2n)^{167}$Tm reactions [10.16]. The radionuclidic purity of the products was poor because 0.7% ^{168}Tm in the ^{167}Tm preparation and 30% ^{153}Tb and 4% ^{156}Tb in the ^{155}Tb preparation were present as impurities. Such a preparation of ^{155}Tb cannot be used in medicine. Preparations of much higher purity can be obtained with the aid of the photonuclear reactions ^{156}D$(\gamma, n)^{155}$Dy/EC/^{155}Tb and ^{168}Yb$(\gamma, n)^{167}$Yb/EC/^{155}Tm, using isotopically enriched targets, as it was shown in [10.11].

Enriched samples of Yb$_2$O$_3$ (20-200 mg) and Dy$_2$O$_3$ (100-1000 mg) were enclosed in aluminium containers and placed just before the tungsten converter irradiated from the other side with an electron beam of 23–25 MeV (beam current 10 μA). The irradiations lasted up to 20 hours. The irradiated targets were left for some optimum time (t_{opt}) to allow the nuclides of interest to grow to maximum activity.

Extraction chromatography on silica gel, pretreated with an organo-silicon compound and coated with a film of di-(2-ethyl hexyl)orthophosphoric acid, was used in ^{155}Tb and ^{167}Tm separation. The Irradiated Dy_2O_3 was dissolved in conc. HCl, the solution obtained was evaporated and the residue was taken up with 1.2M HNO_3. The solution was passed through the extraction chromatographic column and ^{155}Tb was eluated with 1.2M HNO_3. The fractions of the eluate containing ^{155}Tb were evaporated and the residue was taken up with $5-10\,cm^{-3}$ of 0.2M HCl. The enriched Dy was eluated with 3M HCl solution. In the case of ^{167}Tm the solution of the irradiated Yb in 4.3M HNO_3 was passed through the column. The ^{167}Tm was found in the first fractions of the eluate. The recovery of enriched Yb was made with 6M HNO_3. The chemical yield of the separations was $80-90\%$.

Experimental yields of these terbium and thulium isotopes are $3.0 \pm 0.6\,\mu Ci$ per μA per hour and per gram (^{155}Tb) and $2.4 \pm 0.7\,\mu Ci/h\cdot g$ (^{167}Tm). The high purity of the radionuclide preparations should be emphasized. The ^{155}Tb solution contained only ^{157}Tb $< 10^{-2}\%$, $^{155,154?}$Dy $< 10^{-2}\%$, and inactive Dy $< 5\,\mu g/cm^3$. The ^{167}Tm solution contained ^{168,170,172m}Tm$< 10^{-3}\%$, ^{169,175}Yb $< 10^{-2}\%$ and inactive Yb $< 5\mu g/cm^3$. The specific activity of these products was $\sim 0.5\,mCi/g$ $Dy_2(Yb_2)O_3$.

The above results show that ^{155}Tb and ^{167}Tm produced by EA have substantially higher radionuclidic purity than cyclotron products (particularly ^{155}Tb) and are quite suitable for medical use.

10.2.2   Production of Carrier-Free Indium-111

The physical properties of ^{111}In ($T_{1/2} = 67.92\,h$, $E_\gamma = 171\,keV$ — 87.6% and 245 keV — 94.2%) are highly favourable for its application as radiopharmaceutical in the scintigraphic studies. This is a typical 'cyclotron-produced' radionuclide, which was synthesized as early as the fourties. However, for medical purposes it was employed in much later years [10.17, 10.18]. The cyclotron production of ^{111}In involves the irradiation of Ag targets with α-particles [10.19 – 10.21], or cadmium targets with protons and deuterons [10.21 – 10.23]. Short-living ^{109}In ($T_{1/2} = 4.2\,h$) is the main accompanying radionuclide for the reaction. The yields for the proton and deuteron reactions on Cd are significantly higher than those for the α-particle reactions on Ag. In particular, $3-5$ times higher yields of ^{111}In can be obtained at proton (or deuteron) energies of 22 MeV per unit of current, than for Ag targets at α-particle energies of 44 MeV. However, a by-product of long-living ^{114m}In $(T_{1/2} = 49.51\,\mathrm{d})$ is formed during the irradiation of Cd. To reduce this impurity one should use either thin or isotope-enriched Cd targets. It is noteworthy that In can be produced in a variety of ways. In particular, in a cyclotron via ^{112}Sn$(p, 2n)^{111}$Sb(e.c., $\beta^+)^{111}$Sn(e.c., $\beta^+)^{111}$In or ^{110}Cd(^{3}He, $2n)^{111}$Sn(e.c., $\beta^+)^{111}$In reactions [10.24]. Neutron reactions ^{112}Sn$(n, 2n)^{112}$Sn(e.c., $\beta^+)^{111}$In and ^{112}Sn$(n, np)^{111}$Sn at 14.4 MeV in a neutron generator, with cross sections of 1.1 and 0.15 barn, respectively, are also reported.

Malinin *et al.* [10.25] developed the method for the production of carrier-free 111In using bremsstrahlung irradiation. Enriched samples of metallic Sn (150–300 mg) in aluminium containers were irradiated with a bremsstrahlung beam of 25 MeV with a beam intensity of 5-10 μA for 5–10 hours. The bremsstrahlung beam was generated by a running-water-cooled tungsten converter. The isotopic composition of Sn and the impurities of other elements are given in Table 10.3.

The In production can occur either via the ^{112}S$(\gamma, n)^{111}$Sn(e.c., $\beta^+)^{111}$In reaction with a threshold of 11.0 MeV, or through the direct ^{112}Sn$(\gamma, p)^{111}$In reaction with the threshold of 7.8 MeV. Note, that for the bremsstrahlung energy of 25 MeV the yield of the second reaction must not exceed 1%.

Table 10.3. Isotopic composition of the metallic Sn sample and chemical impurities in the enriched Sn samples

Sn isotopes	Content, %	Element	Impurity, mass %
^{112}Sn	80.6	Fe	0.15
^{114}Sn	1.0	Al	0.066
^{115}Sn	0.3	Si	0.1
^{116}Sn	4.6	Cr	0.002
^{117}Sn	1.8	Ni	0.002
^{118}Sn	4.2	Cu	0.018
^{119}Sn	1.6	Pb	0.002
^{120}Sn	4.5	Sb	0.4
^{122}Sn	0.6	Bi	0.028
^{124}Sn	0.8	Zn	0.1

The following separation of carrier-free In was employed. The irradiated Sn was dissolved in concentrated HCl, the acid concentration was brought up to 3M (the Sn concentration to 15 mg/ml), and Sn was extracted with an equal volume of 25%

tributylphosphate solution in heptane. After two successive extractions, the HCl concentration in the aqueous phase was brought up to 2M, and the rest of the Sn was extracted with the same solvent. The aqueous phase containing In was analyzed for the impurities of stable Sn and other elements. The activities of In and other radionuclide by-products were also measured.

The experimental yield of ^{111}Sn for the bremsstrahlung energy 25 MeV in the case of a thin target is $(2 \pm 0.2)\,\mathrm{mCi}/\mu\mathrm{A}\cdot\mathrm{h}\cdot\mathrm{g}^{112}$Sn. The ^{111}In activity in the tin sample (1 g enriched ^{112}Sn$\geq$ 80%) after 10 hours of irradiation, is about $0.13\,\mathrm{mCi}/\mu\mathrm{A}$.

Separation of carrier-free ^{111}In from the irradiated metallic Sn was performed by extraction of the latter from its hydrochloric acid solution with tributylphosphate diluted with heptane.

Analysis of the product showed that the Sn active isotope impurities do not exceed 10^{-4}% in total, while the ^{114m}In by-product content is even less than 10^{-3}%. The content of inactive Sn is less than $1.5\,\mu\mathrm{g}/\mathrm{ml}$, and the content of other inactive elements in total is less than $5\,\mu\mathrm{g}/\mathrm{ml}$.

10.2.3 Molybdenium-99 and Technetium-99m Production

Experience of using the short-living radionuclide ^{99m}Tc ($T_{1/2}$ = 6.02 h, E_γ = 140.5 keV) for diagnosing different human diseases testifies to its high efficiency and innocuousness. Thus, nowadays the problem of a widening its production through the decay

$$^{99}\mathrm{Mo}(T_{1/2} = 66.02\,\mathrm{h}) \xrightarrow{\beta^-} {}^{99m}\mathrm{Tc}$$

is widely discussed.

Nowdays, production of ^{99}Mo are based on the neutron reaction $^{98}\mathrm{Mo}(n,\gamma)^{99}$Mo and fission $^{235}\mathrm{U}(n,f)^{99}$Mo at high-flux nuclear reactors with use of enriched molybdenum (to 99% in ^{98}Mo) or uranium (to 95% in ^{235}U) targets. Irradiation of such targets in reactors provides a high specific yield of ^{99}Mo (up to 10 and 100 Ci/g respectively for irradiation time t_0 =2-7 days and more).

As it was shown by Davydov and Mareskin [10.26], electron accelerators and, in particular, microtrons can also be used for production of these isotopes through the following photonuclear reactions (E_{th} is the threshold of the reaction):

$$^{100}\mathrm{Mo}(\gamma,n)^{99}\mathrm{Mo} \qquad\qquad\qquad\qquad (E_{\mathrm{th}} = 9.1\,\mathrm{MeV})$$
$$^{100}\mathrm{Mo}(\gamma,p)^{99}\mathrm{Nb} \quad (T_{1/2} = 15\,\mathrm{s}) \longrightarrow {}^{99}\mathrm{Mo} \quad (E_{\mathrm{th}} = 16.5\,\mathrm{MeV})$$
$$^{100}\mathrm{Mo}(\gamma,p)^{99m}\mathrm{Nb} \quad (T_{1/2} = 2.6\,\mathrm{m}) \longrightarrow {}^{99}\mathrm{Mo} \quad (E_{\mathrm{th}} = 16.9\,\mathrm{MeV})$$

Simultaneously with ^{99}Mo production a number of long-lived radionuclides of Mo and Nb would be also formed under irradiation of unenriched Mo-target. The parameters of possible photoreactions on molybdenum isotopes are listed in Table 10.4. In the case where corresponding data were missing they were estimated by semiempirical relations [10.27].

Table 10.4. Parameters of the cross sections of photonuclear reactions on Mo isotopes

Nuclear reaction	$T_{1/2}$	E_{th}, MeV	E_m, MeV	σ_m, mb	Γ, MeV
$^{100}Mo(\gamma,n)^{99}Mo$	66.02 d	9.1	17.1	128	7.3
$^{100}Mo(\gamma,p)^{99}Nb$	15 s	16.5	20.4	67	4.7
^{99m}Nb	2.6 m	16.9	22.0	16	4.5
$^{98}Mo(\gamma,p)^{97}Nb$	72 m	15.2	19.8	19	4.8
^{97m}Nb	1 m	15.2	19.8	11	4.8
$^{97}Mo(\gamma,p)^{96}Nb$	23.35 h	14.6	21.9	21	5.9
$^{96}Mo(\gamma,p)^{95}Nb$	35 d	14.7	21.8	21	5.9
$^{95}Mo(\gamma,2n)^{93m}Mo$	6.95 h	17.1	19.6	40	5.1
$^{94}Mo(\gamma,n)^{93m}Mo$	6.95 h	9.7	16.2	185	5.2
$^{92}Mo(\gamma,n)^{91}Mo$	15.5 m	13.1	16.0	195	4.9
$^{92}Mo(\gamma,p)^{91m}Nb$	64 d	12.9	21.0	40	4.9

Here E_{th} is the threshold of the reaction, σ_m is the maximum of the cross section of the giant resonance at the energy E_m, and Γ is the half-width of the resonance.

On the basis of these data the yields of some molybdenum radionuclides were calculated and compared with experiments. These results are presented in Fig. 10.3it a and one can see that the experiment agrees well with calculations.

The calculation of the yield of ^{99}Mo in relation to the total yield of all molybdenum isotopes produced is shown in Fig. 10.3b. It is clear that the relative yield is not practically changed above 24 MeV.

If the irradiation time at 25 MeV equals 100 hours, the accumulation of long-lived impurities ($^{90,91,93}Mo$) would be in the following proportion:

$$^{99m}Tc:^{99}Mo:^{93m}Mo:^{91}Mo:^{91m}Mo:^{90}Mo = 1:5:32:10:6:2.$$

To obtain the maximum accumulation of ^{99m}Tc the irradiated target has to be cooled for 48 hours. By this time the ratio of activity of radionuclides has become:

$$^{99m}Tc:^{99}Mo:^{93m}Mo:^{90}Mo = 100:91:2:0.2.$$

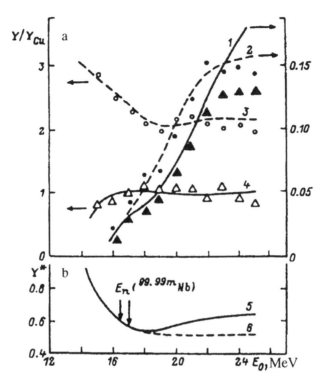

Figure 10.3. a — Relative to copper yields of photonuclear reactions on molybdenum isotopes with electron energy: $1 - {}^{96}$Nb, $2 - {}^{97m+g}$Nb, $3 - {}^{99}$Mo, $4 - {}^{91m}$Mo. Calculations are shown by solid and dashed lines. b — Calculated ratio of the yield of ^{99}Mo to the total yield of all molybdenum isotopes with electron energy: solid line — reaction (γ, n), dashed line — reactions $[(\gamma, n) + (\gamma, p)]$ (from [10.26]).

Successive extractions of ^{99m}Tc from the target have to be done every 24 hours.

In the case of using a natural molybdenum target the latter has to be purified from Nb radionuclides. As it was experimentally shown in [10.26] the complete purification of the irradiated target from Nb took place for targets in the forms $(NH_4)_6Mo_7O_{24} \cdot 4H_2O + H_2O$ (solution) and $(NH_4)_6Mo_7O_{24} \cdot 4H_2O + NH_4OH$ (solution). The ratio of mass of the solution to the mass of molybdenum compound was 1:4. The products of (γ, p)-reactions (niobium) were extracted by their deposition with $Fe(OH)_3$ at pH 12.

The specific yield of the ^{99}Mo is estimated as $(2.4\pm0.3)\mu Ci/\mu A\cdot h$ per 1 g of ^{100}Mo at electron energy 20.5 MeV. This means that to produce $0.25\,Ci$ of ^{99}Mo from a $100\,g$ target with 95% enrichment by ^{100}Mo the necessary irradiation time will be equal 100 hours at average electron beam $25\,\mu A$ (for nonenriched target its mass would be $1\,kg$). Of course, it is not necessary to purify the target from radioactive niobium if the enrichment is rather high. Such a level of accumulation of radioactive ^{99}Mo (and hence of ^{99m}Tc) is sufficient for inspection of $10-30$ patients a day during 1 week (the individual dose for 1 patient is estimated as 10 μCi of ^{99m}Tc).

10.2.4 Production of Iodine-123

In medical diagnosis by inspecting the thyroid gland, kidney and other organs β^-- and γ-emitter $^{131}I(T_{1/2}=8\,d)$ is widely used. For example, in USA every hundredth inhabitant is diagnosed per year, in USSR — every thousandth. Unfortunately, ^{131}I produces rather a large radiative load onto the organism, and therefore it has been gradually substituted by another iodine radioisotope of mass 123, possessing considerably better nuclear characteristics — see Table 10.5.

Table 10.5. Comparative characteristics of ^{123}I and ^{131}I

Characteristic	^{123}I	^{131}I
Half-life, h	13.2	200
γ-ray energy, keV	159	364
β-particles	no	yes
Dose load, rel.un.	1	100

In the late 1980's the production of ^{131}I took place in at least 22 scientific centers such as Karlsruhe (Germany), Vancouver (Canada), Eindhoven (Holland), and Los-Alamos (USA). Using ^{123}I instead of ^{131}I lowers the integral radioactive load onto an organism by about 100 times, allowing the application of iodine diagnosis to children, as in using ^{131}I it is necessary to introduce inside about $1\,mCi$. Short-lived ^{123}I is produced in synchrocyclotrons and cyclotrons with protons of some hundred MeV, but it is necessary to bear in mind, that for medicine one of the most important things is the radionuclide purity of the specimen and its cost. In this relation it seems very attractive to use the photonuclear reaction of xenon $^{124}Xe(\gamma,n)^{123}I$ [10.28, 10.29]. The $^{123}Xe\,(T_{1/2}=2.08\,h)$ produced with 100% probability is transformed to ^{123}I. But it is not so simple in practice as it may seem. Natural xenon consists of nine stable isotopes, and it is only 0.1% ^{124}Xe. So, using natural xenon leads to a final product with a content which is about 4% ^{125}I, formed by the reaction

$$^{126}Xe(\gamma, n)^{125}Xe \rightarrow {}^{125}I \ (T_{1/2} = 60\,d).$$

That is why it is desirable to use xenon enriched with isotope ^{124}Xe.

The threshold of the reaction ^{124}Xe(γ, n) is 10.5 MeV and the maximum of the reaction corresponds to 15 MeV. To exclude other channels of the reaction the electron energy has not to exceed ~ 25 MeV. It has been shown that with the electron energy 22 MeV and with the ^{124}Xe abundance in the target a yield of ^{123}I of about 0.1 mCi/μA·h·g would be achieved.

Figure 10.4 presents the scheme of the installation for ^{123}I production at the microtron in JINR (Dubna) [10.30] The 25 MeV electron beam received from the microtron was focused on the target at 5 m away from the accelerator. A cylindrical tantalum vessel with 2.5 mm walls and 5 cm^3 volume was used as a target. The electron beam interacted with the vessel's bottom, which served as a bremsstrahlung converter. The cylinder was cooled by water. The target center was joined through a needle tap to the communication system, made out of copper tubes 2 mm in diameter. The system consisted of a container with enriched xenon at pressure of about 1 atmosphere, the intermediate 0.25 litre vessel and baths filled with a mixture of liquid nitrogen, alcohol and glycerin (4%). If the correlation of the nitrogen, spirit and glycerin was changed, the temperature would vary from 83 to 173 K in the bath. Previously the system was evacuated up to 10^{-3} torr and then filled with xenon at a pressure of 3 atmosphere. The gaseous xenon was frozen in the intermediate vessel, then gas was pumped to the target by means of the successive dip of the target centre into the liquid nitrogen and alcohol bath. By this method a density of up to 2–3 g/cm^3 could be achieved (the total amount of the substance in the target is about 10–15 g). After the target has been filled it is heated up to room temperature, and this will lead to a pressure increase up to 200 atmosphere.

The high pressure target enriched by ^{124}Xe was irradiated for 10 hours by an electron beam with the intensity of 13 μA, and then the target was cooled down for 2 hours to decay ^{123}Xe to ^{123}I, and the gas was pumped to the intermediate vessel at temperature 170 K. It was found experimentally that all atoms

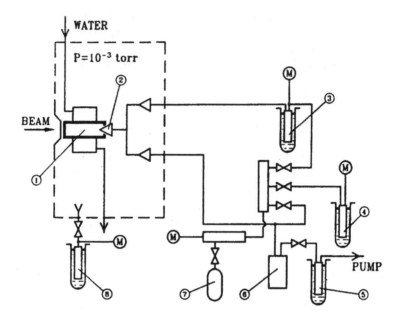

Figure 10.4. The ^{123}I production scheme at the JINR microtron with 22 MeV electron beam: 1 — target, 2 — needle tap, 3 — intermediate vessel, 4 — initial vessel, 5 — nitrogen bath, 6 — zeolite trap, 7 — xenon vessel, 8 — gas collector (from [10.30]).

of ^{123}I with the total activity about 0.2 Ci were kept on the inner walls of the target. Then the target was disconnected from the communication system and carried out to a special room for chemical extraction of the radioiodine. The method makes it possible to produce ^{123}I with the following characteristics:

Productivity (mCi/hour)	20
Iodine concentration in a solution (mCi/ml)	> 200
Radionuclide purity (contaminations)	$< 10^{-6}$
Chemical form (iodide)	$> 95\%$
Solution value (pH)	$7-10$
Content of sodium ions (Mole/l)	10^{-5}
Contamination by heavy metals (μg)	< 0.005

As a result of this work it follows that the ^{123}I of total activity about 50 Ci for year (250 times for 10 hours) can be produced by irradiating a ^{124}Xe target with mass of 10 g. After each irradiation the medical preparation can be made for 1 hour, meeting the guidelines for the pharmacopoeia for direct use in hospitals. As individual diagnostic doses range from $0.1\,\mu Ci$ (the thyroid gland) to $1\,\mu Ci$ (the heart, the liver and so on) and taking into account losses of the preparation's activity before its use, it can be admitted that this amount of ^{123}I is sufficient for diagnosis of over a hundred patients per day annually. Certainly, the specific activity will increase by $8-10$ times, if the electron beam intensity rises to $100-150\,\mu A$, and this is a challenge for present-day accelerator technique.

10.3 Radio-diagnostics

Radio-isotopes are extensively used in radio-diagnostics. The basic mechanism is radio-scintigraphy whereby labelled gamma-emitting pharmaceuticals which concentrate in selected areas of the body can be detected by scintillators. The classical example is the uptake of iodine by the thyroid gland which makes it possible to take a picture of the gland. Conventional nuclear medicine gives only two dimensional projection images. The isotopes which are used may have an appreciable life-time resulting in a non negligible dose to the patient.

Over the last decade a new method of a human body inspection — positron emission tomography (PET) — has come into common use. PET is a method for determining biochemical and physiological process *in vivo* in a quantitative way, including the measurement of pharmacokinetics of labelled drugs and the evaluation of the efficiency of therapy on the metabolism. Positrons, emitted by radioisotopes such as ^{15}O, ^{13}N or ^{11}C which are introduced with suitable compounds into the body, produce, upon annihilation with an electron, a pair of gamma-rays which emerge in opposite directions. The coincident detection of these gamma rays by detectors on opposite sides identifies the straight line passing through the point of annihilation. Tomographic reconstruction determines the distribution of annihilation points from all the coincidence lines recorded. As positrons have a short range, this distribution coincides with that of the decaying nuclei. While conventional radio-chemicals such as iodine can only be used in relation with particular organs, positron emitting isotopes can be used to radio-label water, carbohydrates, proteins, fats or vitamins, i.e. all compounds metabolised by living beings. PET images have better resolution and lower background than single gamma imaging systems.

As we can see, the main positron-active radioisotopes suitable for PET are ^{15}O, ^{13}N, and ^{11}C and they can be produced by (γ, n)-reaction with rather high specific activity. For example, according to the measurements of Oka *et al.* [10.31] their yields under bremsstrahlung irradiation from a 20 MeV linac are 96, 92.5, and 37 MBq/g respectively. Note that there are other radionuclides which are of interest for PET — ^{18}F, ^{30}P, ^{34m}Cl, ^{38}K, and ^{43}K — and they are also can be produced by electron accelerators.

PET has also a growing importance in the diagnosis of cancer. Tumours grow rapidly and require more sugar (i.e. metabolise glucose faster) than normal cells to meet their energy needs. Other recently developed tracers include radio-active amino acids and nucleosides (the constituents of proteins and DNA). These tracers appear capable of distinguishing cancers from benign lesions. PET is also valuable in the diagnosis of epilepsy. It is possible to localise the damaged brain regions responsible for seizures and take corrective surgical action.

The positron emitting isotopes are all short lived and must be produced on site by bombarding targets with protons or deuterons from small cyclotrons. Dedicated machines in the 10 – 20 MeV energy range have been developed for this application. Improved radio-diagnostics will allow a much better localisation of tumours. With classical methods, the accuracy of the tumour configuration established by doctors is much lower than the accuracy permitted by accelerator-based therapy. Close association of better diagnostic tools such as computer tomography, precision delivery systems and accurate patient positioning are expected to improve significantly the efficiency of radiation cancer therapy.

Although most of the accelerator-based diagnostic methods described so far use radio-chemicals, it is worthwhile mentioning a completely different imaging technique based on synchrotron radiation. It relies on the strong absorption threshold for X-rays occurring at certain energies in some atoms. This is the case for iodine at 33 keV. If used as a blood contrasting agent, it is possible by subtracting two images (taken with X-rays below and above the threshold) to obtain images of the patient's vascular system and diagnose coronary arteriosclerosis which may result in a deadly blockage of the oxygen supply to the heart muscle. Although the principle of this method was proposed over 40 years ago, it has only recently become practicable with the tuneable monochromatic X-rays produced by electron synchrotrons.

Chapter 11

Therapeutic Microtrons

Radiation therapy is one of the main methods for curing cancer. The number of people attacked by cancer is enormously high. According to the World Health Organization estimates, currently there are approximately nine million new cancer cases per year, worldwide. This number is expected to increase to about 15 million new cases by the year 2015, with about two-thirds of these cases in developing countries. Therefore medical applications of accelerators as sources of fast particles, developing irradiation techniques and wide investigations of biological aspects of irradiation are some of the most noble tasks of modern nuclear physics.

11.1 Radiation Therapy

The main aim of radiotherapy is to destroy affected cells in such a way as to prevent further spreading of the tumour. We know from experience that ionizing radiation in the patient will damage certain malignant tumours more than the normal tissues in which these tumours grow and this fact makes it possible to destroy such a tumour without irreparably damaging the normal structures. When the therapeutic ratio is large we can encompass considerable portions of the body and get a high cure rate even if there is extensive spread. But in general this is not the case, and the tolerance of the normal tissue is the limiting factor in applying the dose of radiation. As a rule, it is impossible to irradiate a tumour without irradiating some normal tissue and it is accepted that the greater the volume of normal tissue irradiated, the greater the damage it will suffer for a given dose. That is why now intraoperational irradiation is also performed, in general by high-energy electron or photon beam.

295

11.1.1   Physical Considerations

Both kinds of radiation from electron accelerators — high-energy electrons and bremsstrahlung — have been used for radiation therapy since the early fifties.

When high energy electrons pass through matter, most of the electrons come to stop near the end of the range. Electrons, being light, travel with relativistic speed through most of their range; hence, the dose deposited by them, as a function of depth, remains nearly constant and then decreases rapidly near the end of their range. Figure 11.1 illustrates this statement.

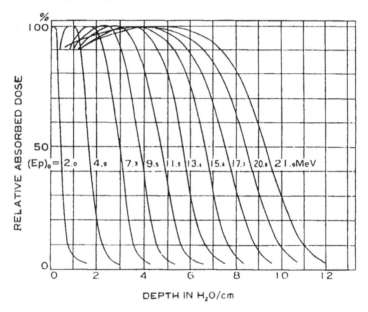

Figure 11.1. Electron depth dose curves; field size 12×12 cm, $(EP)_0$ is the most probable energy at the phantom surface (from [11.1]).

As for γ-radiation, it is well known that γ-rays, as they pass through the medium, are attenuated due to photoelectric, Compton, and pair production processes. Hence, the dose delivered by them decreases exponentially with depth except for initial build-up. The build-up is so that we get a dose which is vary small on the surface, rises quickly to a maximum and the depth of the maximum is a function of the gamma-ray energy; the higher the energy, the deepest the maximum, as can be seen from Fig. 11.2.

This behaviour determines one of the advantages of high energy radiation as compared with conventional radiation, that is, for the same tumour dose, the integral dose is smaller.

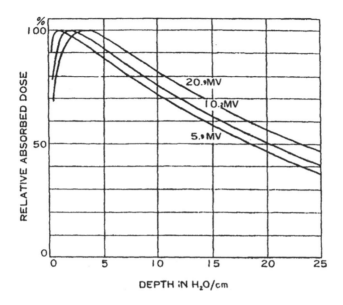

Figure 11.2. Gamma-ray depth dose curves; field size 25 × 25 cm (from [11.1]).

11.1.2 Biological Considerations

Looking for the biological effect, the term LET (linear energy transfer) for any kind of radiation is convenient when the oxygen effect is considered. It is known that many tumours have inadequate blood supply, hence, they contain hypoxic cells.

For conventional radiation, the dose required to sterilize hypoxic cells is about three times that for oxygenated cells. The presence of hypoxic cells in the tumour therefore requires an increase in the tumouricidal dose. Hence, when low LET radiations are used, any further developments in dose distribution cannot avoid damage to normal tissues that are within the tumour volume.

The radiation resistance of hypoxic cells when compared with oxygenated cells, is reduced with increasing LET. Thus, high LET radiations may be more effective in overcoming the hypoxic cell problem where the current methods are not successful.

However, it is important to outline that for any kind of radiation we should distinguish two components of LET. For instance, the density of ions for the α-particles is 100 to 1000 times greater than for high-energy electrons. However, secondary and tertiary electrons produced by the primary particles will change this situation [11.2]. The fact that we now get a spectral distribution of electron energies and low energy electrons which may produce 100 times denser ionization than the high-energy electrons has to be recognized. In the spectrum of LET we can therefore distinguish between high-LET and low-LET parts. When D is the dose, the high-LET component will be αD, where α is a constant typical for the type of radiation, while the rest of the dose, $(1 - \alpha)D$ will represent the low-LET component.

When discussing the influence of different radiation types (i.e., different values for the factor α) on radiation therapy, we have to compare the two radiation effects separately. Regarding the high-LET component effects, different cell types show fairly similar behaviour — normal and tumour cells might have nearly the same sensitivity. For the low-LET components, however, the differences are much greater, the normal cells being 3 to 5 times less sensitive than tumour cells [11.2].

This fact, based on clinical observations, might be due to the higher differentiation of normal cells as compared with proliferating tumour cells. Therefore, in order to improve the selective destructions of tumour cells (i.e., obtain a high 'effectivity') the low-LET component should be mainly brought into action. This means that the high-LET component should be small, indicating the use of high-energy electrons, and also that high single doses should be given. The magnitude of the single dose will, of course, be restricted because of the tolerance limit of normal tissue reactions, and especially damage to the vascular system. The fact that high-LET radiation could contain low-LET component radiation, turns out to be fortunate for radiotherapeutic purposes.

From the work carried out by Barendsen [11.3] using α-particles of different energies, the following results evolved regarding the survival curves for single and fractionated exposures for 250 keV X-rays and 3.4 MeV α-particles. Because of the capacity of the cells to recover in the shoulder region (coinciding with the break points of fractionation) — see Fig. 11.3, fractionated X-ray doses are less effective than single doses. However, for high-LET as for 3.4 MeV α-particles (in this case, the survival curves are exponential), all the damage done is lethal. Thus, the fractionated doses of high-LET particles are just as effective as single doses.

The very steep dose response characteristics often observed in tumours and surrounding healthy tissue, imply that a high precision in the absorbed dose and a high degree of dose uniformity

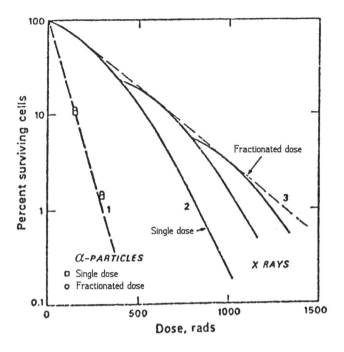

Figure 11.3. Surving curves vs dose

is of primary importance in accurate radiation therapy. These factors are therefore decisive in the design of high quality therapy beams.

11.1.3 Optimisation of Treatment Fields

As we emphasized, high energy photons cause low surface doses but penetrate deep into matter, while momoenergetic electrons show the opposite effect, being directly ionising and having a rather well defined range. Both kinds of radiation must be available for effective therapy, and often, broad homogeneous treatment fields with smallest possible penumbra (steepest possible edges) are desired, and for that reason the fields must be flattened and collimated. Due to the dissimilar physical characteristics of electrons and photons, different strategies must be applied to bring about the desired field in the two cases, as was considered by Rosander [11.4].

In electron therapy, it is generally of great importance that the treatment fields are formed in such a way that the good energy homogeneity of the microtron beam is preserved. Then the terapeutic range is at the maximum for a given beam energy and, beyond this range, the absorbed dose has the steepest possible decrease. To accomplish this, the amount of scattering material is minimised by the use of dual scattering foils, a thin primary foil of high Z material and a shaped aluminum secondary foil in the central part of the beam. In fact the air is also used as a scatterer, giving a lateral contribution to the field from the region surrounding its geometrical edge, as only the foremost part of the adjustable electron collimator defines this edge. By these measures very uniform therapy increases the surface dose. The energy of this new field component is low enough not to affect appreciably the fall-off part of the depth dose distribution.

The intensity and photon energy of the unprocessed photon beam from a bremsstrahlung target is markedly peaked in the forward direction. When thick low Z flattening filters are used, the high average energy of the photons in the central part of the beam is maintained or even increased and, due to the angular distribution of the photon energies, good flattening can be obtained for a particular depth. In medical microtrons another solution to the dose flattening problem is applied. The flattening filters are made composite, consisting of a central high Z part (lead) placed in a low Z cup (aluminium). The central part is egg-shaped, so that the effective atomic number is gradually changing from the one to the other along a radius. Thus, apart from flattening the intensity, such a filter also reduces the mean photon energy in the central part of the field, while in the periphery, the average energy is increased. This action depends on the high probability of pair production by high energy photons in heavy materials and of photoelectric absorption in light elements at low energies. At the cost of somewhat reduced penetration, the method results in broad therapy being flat at all depths.

11.2 Medical Microtrons

For radiation therapy at medium and high energies the circular and race-track microtrons are ideally suited electron accelerators. The fine energy definition of the electron beam from a microtron allows the use of beam transport from one accelerator to several treatment units. This possibility makes an installation with two or more treatment rooms very interesting from an economical point of view, particularly at high energies.

The microtrons for cancer treatment are designed and produced in two places — in Moscow by the R&D 'AGAT' Corporation, and in Sweden by Scanditronix. The first medical microtron was installed by Scanditronix in 1974 at the Spedali Civili hospital at Brescia in Italy, and since then about 20 circular medical microtrons with 10, 14, and, most often, 22 MeV maximum beam energy have been fabricated and installed by the company. In 1984 a 50 MeV racetrack therapy machine was comissioned in Umea and most accelerators of this type are now developed and produced in Sweden.

The 'AGAT' Corporation in Russia has been developing effective circular medical microtrons since the late 70's. The pilot model of the medical microtron at energy 22 MeV was installed in the Hertzen Research Institute of Oncology in Moscow. It has been used for both photon and electron cancer treatment since 1985, and recently (1997) more than 10000 patients have been treated. Industrial manufacturing medical microtrons are also installed in the Medical Radiological Institute (Obninsk, 1993), in the Institute of Oncology and Radiology (Minsk, 1993), in the Estonian Cancer Centre (Tallinn, 1993), in the Ukranian Cancer Centre (Kiev, 1996), and in the Moscow Region Clinical Research Institute (1997).

Below we will consider in detail the MT22 Russian circular medical microtron and the race-track medical microtron MM50 of Scanditronix.

11.2.1 Russian Circular Medical Microtron

The development of medical microtrons in the former Soviet Union dates back to 1970–71 when the scientific group from AGAT Corporation, leaded by Mirzoyan, attempted to install the 22 MeV microtron on the existing stand of the therapeutic betatron in one of the Moscow clinics. In 1976 the development of therapeutic units with a microtron in the isocentrical version were initiated and in 1982 the first model of the 'Microtron-M' system was installed in the P.A.Hertzen Moscow Oncological Institute where in 1984 the clinical tests were completed.

11.2.1.1 Therapeutic Unit

The 'Microtron-M' system includes the following separate parts: a therapeutic unit with a microtron, a pulse modulator, a control racks, a control desk, a therapeutic table, a cooling system and a rough-vacuum pumping system. The power supplies for separate components of the control system, signalling, interlocking and the safety system are placed in the control racks.

The main characteristics of the microtron M22 are presented in Table 11.1.

Table 11.1. Russian medical circular microtron M22

Number of orbits	22
Max. output electron energy, MeV	22
Magnetic field, T	0.2
Magnet diameter, m	1.27
RF pulse power, MW	1.6
RF frequency, GHz	2.8
Pulse width, μs	3.0
Nominal electron energies	8, 12, 15
for cancer treatment, MeV	18, 20, 22
Max. dose rate, Gy/s:	
electrons	0.083
photons	0.05
Max. photon energy, MV	20
Max. field size	
at the isocentre, cm^2:	
electrons	20 × 20
photons	30 × 30
AC power, kW	26
Cooling water, l/min	20

The selection of irradiation modes required as well as starting and stopping the ionizing radiation are performed from the control desk. The dose monitors, providing the setting of irradiation dose required, the control and the indication of dose delivered and the automatic termination of radiation when a set dose and a delivered dose coincide, are also placed in the control desk. The functions of the other devices are well known and we consider only the main units in detail.

The therapeutic unit is the main component of the 'Microtron-M' system (Fig. 11.4). It consists of a movable part (a gantry which is a support structure for the treatment head and accompanying beam transport components) and a stationary one. In the stationary part there are located the rotation drive, the pulsed transformer, the pulsed magnetron with the magnetic system and the device for magnetron frequency tuning. The stationary part includes the waveguide assembly with the rotary waveguide joint and the force axle that is a drum resting upon two bearings.

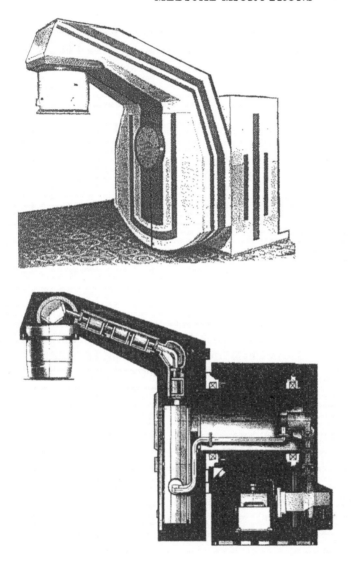

Figure 11.4. Photo and schematic drawing of inner parts of the medical microtron M22

The rotary gantry is a welded steel frame of a complicated shape with the console at an angle of 20 degrees to the axis of rotation, which is rigidly fastened to the force axle. The microtron (in vertical median plane position), the part of the waveguide circuit with the T-circulator as a decoupling element, the electron beam transport system and the therapeutic head are placed in the flat vertical part of the gantry. The front part of the gantry is divided into two movable doors providing access to the microtron. The isocentre height is equal to 130 cm, the total angle of rotation of the gantry equals 360 degrees.

11.2.1.2   Microtron and Beam Transport System

The magnet poles in a microtron are flat and the requirements for field uniformity depending on the number of orbits are rather high. In the microtron under consideration the nonuniformity of the magnetic field does not exceed 0.2–0.3% even on the last, 22nd orbit.

The extraction channel is a conical steel tube 70 cm long. The outer diameter of the entrance of the tube is equal to 16 mm and of the exit of the tube is 40 mm, the inner diameters are equal to 12 and 30 mm respectively. The magnetic field disturbance on the next to last orbit caused by the tube is completely compensated by two solid cylindrical iron rods located symmetrically above and below the median plane.

The vacuum rings of the vacuum chamber are made of stainless steel. In the latest model of the microtron the magnet yoke is used as the vacuum ring. The accelerating cavity is excited with the magnetron connected to the cavity through the waveguide circuit. The waveguide adjacent to the cavity is placed in the vacuum. A quartz window separates this part from the other parts of the waveguide circuit filled with freon. The ferrite T-circulator placed between the magnetron and the cavity ensures reliable decoupling (25 dB and more). Automatic frequency tuning is used in the system to maintain stable operation of the cavity (see Ch. 2).

The accelerating gap and the coordinate of the cathode (in dimensionless units) in the cylindrical microtron cavity are equal to $l=1.45$ and $x_o=0.2$ respectively. The microtron operates in the second type of acceleration. The value of the magnetic field in the units of cyclotron field is equal to 1.85, and the number of orbits is 22. The flight apertures in the cavity are radial slits. When the microtron operates in the second type of acceleration the operating conditions of the cavity are rather hard. For example, the amplitude of the rf electric field reaches about 1 MV/cm, that is why high requirements are imposed upon the mechanical treatment of the inner surfaces of the cavity. A cylindrical rod 3 mm in diameter, 5 mm long and made of lanthanum hexaboride is used as a cathode. The cathode is fastened to a thin tantalum holder which ensures both a fairly reliable electric contact and good fixation of the position of the cathode. The cathode is heated with a subsidiary electron beam from the heated tungsten spiral filament. A potential of 500-600 V is applied between the cathode and the filament. The power of electron heating of the cathode is equal to 25 W. The emission current can be smoothly controlled by changing the heating power.

A vacuum of 10^{-4} Pa in the microtron vacuum chamber and in the beam transport system which is required for a normal operation, is achieved through continuous pumping with the magnetic discharging pump. Its blocks are installed inside the vacuum chamber away from the median plane. Preliminary pumping of the vacuum volume is performed by the rough-vacuum pumping system.

The accelerated electrons are extracted from the last orbit through the magnetic channel and then through the beam transport system of a special construction consisting of two bending magnets and four quadrupole lenses entering the therapeutic head, as seen in the schematic drawing of the M22 in Fig. 11.4. The beam transport system ensures beam passing and focusing onto the photon target practically without loss. The diameter of the focused electron beam is about 3 mm.

The formation of irradiation fields, as well as patient protection against useless radiation, are provided by elements of the therapeutic head. The therapeutic head incorporates the target for photon producing, a set of primary scattering foils, a primary collimator, flattening and compensating filters, an ionization chamber, diaphragms, wedge filters, a light system for simulation of the irradiation field and the optical range-finder. The thickness of the target equals one-third of the radiation length of 20 MeV electron in tungsten. The composite filter is used to flatten the field intensity of photon irradiation. The profile of the filter should be corrected by experiment. A so-called two-foil system is used to form the uniform electron irradiation fields [11.5].

The primary scattering foils in the shape of lead and graphite discs are placed at the entrance of the movable part of the collimator. In addition to its direct function of increasing the angular divergence of the electron beam, the graphite discs also serve as moderators, in which the primary 22 MeV electron energy loses up to 14 MeV (depending on the thickness of the disc). The primary scattering foil for 22 MeV energy is made of 0.1 mm lead foil. The preliminary scattered electron beam falls onto the aluminium compensator. The ionization chamber serves to obtain information about the passing radiation and consists of four small chambers, their signals being processed independently. Two chambers with electrodes, which completely overlap the beam cross section, form two independent channels to measure the delivered dose. Two other chambers are used to determine the position of the beam. Their electrodes are made on the same plate in the shape of strips, which are located at right angles to each other. The difference signal derived from these chambers, after appropriate processing, enters the power supply of the bending magnets for the automatic space stabilization of the beam position.

The therapeutic head described above makes it possible to obtain both photon and electron irradiation fields which correspond to the requirements of the International Electrical Commission (Secretariat, 62C) for medical electron accelerators with $1-50$ MeV energy [11.6]. Figure 11.5 illustrates the profiles of relative dose distributions for electrons and photons in the direction perpendicular to the radiation beam.

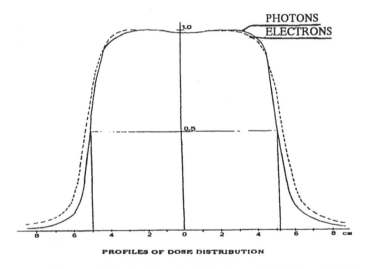

PROFILES OF DOSE DISTRIBUTION

Figure 11.5. Profiles of relative dose distribution at M22.

The irradiation of a patient with various types of radiation or with electrons of various energies can be performed during one session with the patient. The change of the type of radiation is performed by installing an electron applicator weighing 3 kg and this process takes 2 minutes. The time of changing the electron energy during the electron therapy does not exceed 40 s.

Figure 11.6 illustrates what types of cancer treatment were carried out in the Hertzen Oncological Institute during the period from July 1985 to December 1993. The total number of patients which were treated by the 'Microtron-M' therapeutic system was equal to 5103 (as mentioned, by 1997 this number was over 10 000). They received 87 703 sessions (60 treatment sessions per day in average), and one-third of the patients were treated by combined photon and electron radiation. Intraoperational irradiation was also performed on 105 patients with the use of an electron beam at 6, 12, and 18 MeV..

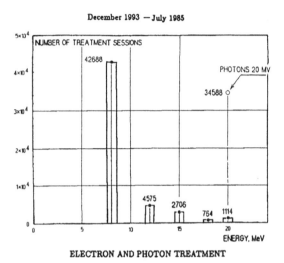

Figure 11.6. Total number of patients treated by electron and photons of different energies (courtesy of A.R.Mirzoyan).

11.2.2 Race-Track Medical Microtron MM50 of Scanditronix

The radiation modalities of most frequent use today are of low and medium energy, from a few MeV up to about 20 MeV. This is the case even though, for many target locations, better dose distributions are obtained at considerably increased beam energies. The development and use of high energy radiation modalities with energies up to about 50 MeV have been limited by a number of factors.

One of the most important reasons is the technical problems of obtaining radiation beams of high quality at these high energies. In the case of electron beams the advantageous steep dose fall-off behind the therapeutic range is largely lost, in particular when ordinary scattering foils are used to flatten the beam. For the photon beams the problem is instead that the very thick flattening filters needed will decrease the effective photon energy, so that only a small improvement in beam penetration is obtained by increasing the beam energy.

The development of race-track microtrons during the 70's at two research institutes in Sweden, the Royal Institute of Technology in Stockholm' and the Department of Physics, Lund University made it possible to design the clinical 50 MeV race-track microtron [11.7] which has been in use for regular patient treatments since the summer of 1988. In this machine a number of new and effective treatment procedures were introduced, as considered by Rosander [11.4].

Few tumours have a rectangular shape. Straight collimator jaws, although individually adjustable, can only very approximately imitate the projection of a tumour volume and, thus, sparing nearby healthy tissue, often more sensitive to radiation than the oxygen-weak tumour, is difficult. This problem is greatly relaxed by the so called multi-leaf collimator, which consists of two equal groups of together 64 tightly packed, parallel tungsten plates, whose position is individually adjustable from a distant position on one side to fourteen centimetres on the other side of the gantry axis. In this way, irregularly shaped therapy fields of any desired cross section within $32 \times 40 \, cm^2$ can be created. In order to avoid excessive penumbras, all sides of the plates point towards the scattering foil/photon target, and the movement of the plates follows a circular arc, centred at this same point.

Another interesting part of modern high energy medical microtrons is the beam sweeping system, which gives a new dimension to therapy as, now, a narrow beam can be swept over the target volume in an arbitrarily complicated pattern to create almost any distribution of the treatment field. The scanning system consists of a 97° magnet — the final bending magnet of the gantry — on either side equipped with a sweeping magnet (Fig. 11.7).

The bending magnet has edges slanted for double focusing and images the centre of the first sweeping magnet onto the centre of the second. The image distance is made short to create a small source. The first sweeper scans the beam in the bending plane of the 97° magnet, and the second sweeper acts in the perpendicular plane. Thus, the outgoing beam, being swept in two dimensions, will always come from a point in the centre of the second sweeping magnet. When photons are wanted the target is placed immediately downstream from this magnet. To avoid contamination of the photon fields by electrons and positrons originating from the radiator, the swept photon beam is filtered by a purging magnet. The gantries are also helium filled in order

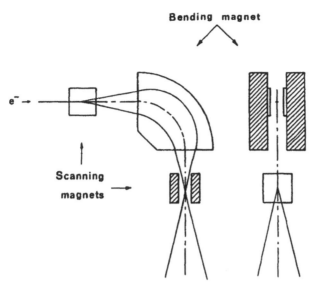

Figure 11.7. Design of scanning system for the electron and photon beams from a microtron (two orthogonal projections). The photon beam is scanned by placing the target immediately below the last scanning magnet (from [11.7]).

to reduce corruption of the fields due to electron contamination in the photon case, and blurring of the beam due to air scattering in the electron case. In this way the electron fields, as an example, can be so exact that applicators (masks placed on the patient) can be avoided. This is, of course, a prerequisite for arc (moving source) therapy with electron beams.

The race-track microtron shown in Fig. 11.8a produces electron beams with energies ranging from 5 to 50 MeV in steps of about 5 MeV. The electron beam from this unit may be handled in one or more gantries where both scanned electron and scanned photon beams can be generated [11.9, 11.10]. The principal view of the therapeutic head is shown in Fig. 11.8b. The scanning system was designed to keep the effective source as small and well-defined as possible. To achieve this the electron beam is scanned in such a way that it will always pass through the same geometrical position. This scanning system is used with different types of scanning matrices for both electron and photon beams.

The main collimating system (1) of this unit is a double-focused multileaf collimator with 64 leaves, each with a width of 1.25 cm at the isocentre. The maximum field size at the isocentre

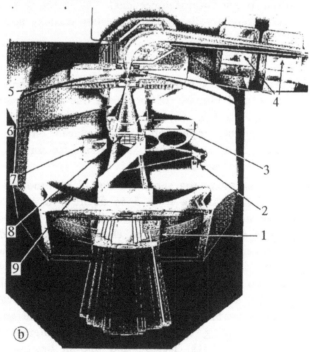

Figure 11.8. *a* — 50 MeV race-track microtron for therapeutic use (from [11.9]), *b* — principal view of therapeutic head (from [11.10]).

is 32 by 40 cm and the possible cross-centre movement of the leaves is 5 cm at the SSD (source to surface distance)=100 cm. The thickness of the tungsten leaves is 8 cm. A video camera and semitransparent mirror (2) together give a beam's eye view of the collimator setting for real-time verification. A five position filter revolver (3) primarily acts as a collimator, and can be fitted with filters for special treatment techniques. The seven position target exchanger (5) is equipped with optimized bremsstrahlung targets and electron decelerators. The purging magnet (6) which acts as a primary collimator, is placed after the bremsstrahlung targets. This magnet cleans the photon beams from contaminating electrons and positrons to minimize surface dose and maximize the build-up depth. The built-in automatic dual wedge filter mechanism (7) prevents set-up error and the time-consuming handling of heavy accessories. Monitoring of the intensity distribution of the scanned beam is performed by the transmission ion chamber (8). The therapeutic head is filled with helium (9) to minimize electron scattering to allow accurate multileaf collimation, and reduce the electron contamination of the photon beam.

The electron scanning of the MM50 is performed in a hexagonal pattern (Fig. 11.9). After completion of one scan fraction the positions of the scan points are displaced in order to give uniform beams. The scan pattern is divided into four 'scan fractions', each covering the field size chosen. The positions of the scan fractions are displaced to achieve a more dense scan pattern and hence a more homogeneous field.

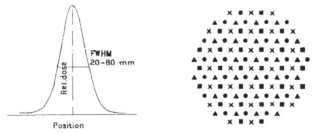

Figure 11.9. Example of elementary beam and scan pattern for electrons; *, •, △, □ indicate the different scan fractions (from[11.9]).

As it was noted by Rosander [11.4], to take full advantage of the beam sweeping technique, it is of great value being able to vary the pulse dose from one pulse to the next. As therapy accelerators generally are operated at high repetition rates, the intensity variation must be quick, and this is probably most easily done by the aid of a grid-equipped electron gun. Gridded electron guns can also be used for the injection of nanosecond electron beams into accelerators. It is believed that intensive pulses of a duration shorter than, say, 10 ns might be advantageous for therapy, as then the majority of the free radicals, created in the patient, would be simultaneously present.

Chapter 12

Industrial Applications

The microtron, being a powerful and efficient source of radiation, can be successfully used not only for scientific research but also to solve various technological problems such as material detection and inspection, isotope production, and as radiation processing, considered in this chapter.

12.1 Industrial Radiography

Developments in the field of heavy industry, manufacturing high temperature and high pressure vessels, high-power steam turbines, heavy-duty transport vehicles, and other components for classical and nuclear power plants and chemical installations, call for stringent requirements concerning nondestructive testing.

Quality inspection is therefore needed to ensure the security of operating the equipment mentioned. We must face the fact that microtron inspection is not the only answer. On the international level, there is a strong competition among ultrasonic inspection, gamma-radiography (using vary large cobalt-60 sources), and radiography using linear electron accelerators.

Radiography has several well known advantages over ultrasonic inspection, especially when thick-walled castings of complicated shape and rough surfaces have to be tested. Besides, multilayer pressure vessels cannot be inspected ultrasonically.

The main advantage the microtron has, as compared with kilocurie cobalt sources, are shorter exposure times, higher quality radiographic images, enhanced inherent safety and the lack of source replacement necessity.

Nondestructive testing is of special interest for use in nuclear power plants. The need for maintenance in nuclear reactors is

313

reduced to a minimum in two ways: first, by use of the best and most corrosion-resistant materials even at higher costs; and second, by use of the best possible manufacturing techniques and procedures. Nondestructive testing is used as an adjunct to the manufacturing processes, to reduce the numbers of component failures and leaks in nuclear reactors to the absolute minimum.

The radiographic sensitivity and resolution requirements for inspection of nuclear components are generally more severe than those accepted in most industrial practices. Since radiography of a great range of materials and thicknesses may be necessary, a broad range of gamma-ray energies must be available. This requirement can be easily accomplished by a variable energy microtron.

The majority of scientific and technical tasks are to find out inner latent defects and structural distortions in different objects On the other hand, current technology makes it possible to control their inner condition over time without their destruction. This field of radioscopy is named dynamical radiography and it is directed to the study of the internal structure of an object during its actual operation. For instance, a simultaneous neutron and gamma radiography station is in operation at a steady state research reactor in Hungary [12.1]. Neutrons and gamma radiation passing through the object are converted into light by means of neutron- and gamma-sensitive scintillation materials, respectively. The light is detected by a low-light-level TV camera. The imaging cycle of the camera is 40 msec, thereby providing the possibility for visualizing medium speed movements inside the investigated object. The radiography images are displayed on a monitor and stored by video recorder for further quantitative data analysis. This radiography station has been used for many years for inspecting industrial products, with the aim of research and development.

12.1.1 Special Microtron for Industrial Nondestructive Testing

A special compact variant of a circular microtron was designed and manufactured in Russia during the 80's for the needs of nondestructive testing [12.2]. The main feature of this microtron is that the waveguide, ferrite insulator and magnetron are placed on the electromagnet pole. This was achieved by using resonance waveguide junctions and an electromagnet instead of a permanent magnet for the magnetron. Inside the stainless steel vacuum chamber there is a magneto-discharged titanium pump that also decreases the size of the installation. The vacuum part is separated from the magnetron by a quartz window which is placed in a waveguide. The microtron accelerates electrons up to 12 MeV and its dimensions are 1.1 × 1.1 × 1.5 m and 1500 kg weight.

A bremsstrahlung tungsten target (0.7 mm thick) and tungsten collimator are placed inside the electromagnet. The effective diameter of the electron spot on the target equals 3 mm. At the first type of acceleration the final electron energy in this machine equals 8 MeV and the radiation dose is 1000 R/m. The sensitivity of the radiative control of steel products 70–300 mm thickness is equal to 0.6–0.9%. Fig. 12.1 *a,b* shows the dependence of the sensitivity of radiative control and the exposure time on the thickness of the product.

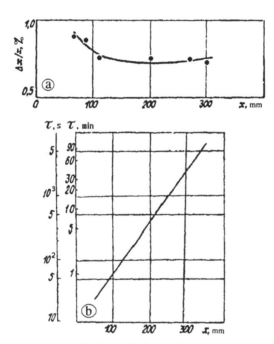

Figure 12.1. *a* — Sensitivity of the radiative control depending on the thickness of the product; *b* — exposure time depending on steel products (from [12.2]).

12.1.2 Stereomicrotron

A stereomicrotron is a microtron in which one may alternately accelerate two electron beams [12.3]. The stereomicrotron may find application in industry for stereoradiography of very thick parts for the purpose of determining the depth of a fault in the part and determining its shape.

Figure 12.2 displays the overall diagram of a stereomicrotron. Two emitters 2 and 3 are installed on the side walls at identical distances from the resonator axis in the accelerating resonator 1 which is excited from the oscillator 10. The transit apertures in the resonator are made of identical size. Such a resonator shall henceforth be called symmetrical. Two stopping targets 4 and 5 with radiation collimators 6 and 7 made of a material with a high atomic number are installed symmetrically relative to the common diameter of the orbit on the last electron orbit. Using a mechanism situated in the vacuum chamber, one may move the target with the collimators remotely along the last orbit. This is necessary for the longitudinal axis of the collimator to coincide with the tangent to the electron trajectory.

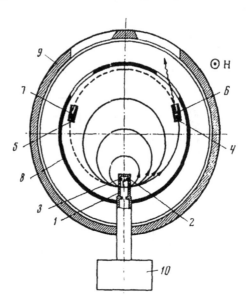

Figure 12.2. Overall diagram of the stereomicrotron (from [12.3]).

When microwave power is applied to the resonator and a magnetic field with the direction indicated in Fig. 12.2 is turned on, the conventional acceleration process takes place, the electrons being captured solely from emitter 2. The γ-radiation which develops during the stopping of the electrons in target 4 exits from the accelerator through special windows in the vacuum chamber 8 and in the return magnetic core 9. When the direction of the magnetic field changes, the acceleration process repeats, only now emitter 3 and target 5 are used, while the electrons move along their orbits in the opposite direction. By moving the target along the orbit, one may vary the angle of intersection of the bremsstrahlung beam outside the accelerator (i.e., one can choose the conditions for obtaining optimal stereoscopic vision).

The principle of operation of the stereomicrotron was checked using the microtron at the Institute for Physical Problems (Moscow) which has 30 orbits. A cylindrical resonator 1 of the type with acceleration in an efficient regime was used in the experiments. The dimensions of the transit aperture were: vertical 0.085λ and radial 0.23λ (here λ is the wavelength of the accelerating rf field). The centres of the aperture coincided with the resonator axis. When the experiments were carried out with one beam, one emitter made of tantalum with an emitting surface $0.03 \times 0.03\lambda^2$ was installed on the resonator. The emitter was heated with direct current. The measurements were carried out on the twelfth orbit where the energy of the accelerated electrons amounts to $\simeq 8\,\text{MeV}$.

Figure 12.3 displays the results of measuring the dependence of the capture coefficient on the parameter Ω, whose range of variation corresponds to the variation of the electron energy on the twelfth orbit in the $6.5-9\,\text{MeV}$ range.

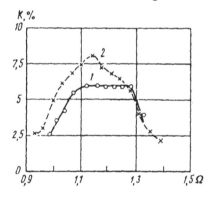

Figure 12.3. Dependence of the capture coefficient on the variation of the magnetic field: 1) symmetrical resonator, 2) resonator ensuring the best vertical focusing (from [12.3]).

Curve 1 was obtained for the configuration of apertures indicated above. The maximum capture coefficient under these conditions equals 6% for $\Omega = 1.12 \div 1.28$. Curve 2 illustrates the capture coefficient for the case when the choice of the dimensions and position of the transit apertures ensures optimal vertical focusing of the particles during acceleration. The entry aperture has the same dimensions and position as in the preceding case, while the radial dimension of the exit aperture is twice as small. The centre of this aperture is shifted along the radius from the resonator axis by 0.03λ toward the emitter. It follows from Fig. 12.3 that in this case the maximum capture coefficient is equal to 8% for $\Omega = 1.14$.

Since the parameters of both beams in a stereomicrotron must be identical, the resonator must lie symmetrical, notwithstanding the fact that the losses of accelerated particles in the resonator due to a weakening of the vertical focusing is greater than in the resonator of a conventional microtron.

The process of reaching a steady-state acceleration regime for a change in direction of the magnetic field was likewise investigated in order to determine the maximum frequency of magnetic-field switching. The direction of the field was changed by switching the direction of the current in the excitation windings of the electromagnet. The current of the electromagnet and the average current of the accelerated beam on the twelfth orbit were recorded simultaneously. It was clarified that for an increase of the magnetic field the acceleration in the microtron begins when the magnetic field differs by approximately 10% from the equilibrium value. Such a field corresponds to the equilibrium phase on the boundary of the phase-stability domain.

As experiments showed, after the direction of the magnetic field was changed, the accelerated beam appeared after the time $\sim 2.3\tau$ had elapsed (τ is the time constant of the electromagnet), and its full value was restored after $\sim 5\tau$. Thus, the maximum field-switching frequency in a stereomicrotron with a solid magnet must not exceed $f \simeq 1/5\tau$ and is determined by the time constant of the electromagnet rather than by the action of the eddy currents, since the rate of change of the magnetic field is negligible after the time 5τ has elapsed.

12.2 A Microtron-Based Detector for Explosives and Narcotics

Detection and imaging of objects rich in nitrogen, such as modern explosives or bulk narcotics, is one of the high priority present-day tasks. Control of airline baggage, inspection of vehicles and cargo containers, and detection of buried land mines are obvious applications. The basic technique, proposed by Luis Alvarez as it was noted in [12.4] and developed by Trower with coworkers [12.4, 12.5], has two major components: the comparatively unique response to high-energy photons of the nitrogen nucleus and the compactness and energy variability of the racetrack microtron. The project of a mobile 70 MeV race-track microtron, which is specially designed for solving this task, was considered in Chapter 3. Here we discuss the nitrogen camera which produces elemental images for concentrations and surface densities typical of concealed bulk narcotics and terrorist explosives.

The nitrogen camera is an instrument based on a nuclear technique which is capable of imaging nitrogen concentrations with surface densities and amounts typical of currently concealed conventional explosives. The main idea is to scan an inspected object by high-energy bremsstrahlung and to detect short-living radiation which is resulting from photoreactions on nitrogen. Experiments with a nitrogen camera [12.5, 12.6] were performed with 50 MeV bremsstrahlung from a 50 MeV race track microtron of the Royal Institute of Technology in Stockholm.

Among the variety of radioisotopes which are produced under high-energy bremsstrahlung irradiation of an object only two are of interest, i.e. can serve as nitrogen markers, and they are ^{12}N and ^{12}B. These nuclides disintegrate with half-lives and end point energies of $11.0 \, ms / 17.3 \, MeV \, (e^+)$ and $20.2 \, ms / 13.3 \, MeV \, (e^-)$, respectively.

Three reactions produce these two radioisotopes:

$$\gamma + {}^{14}N \rightarrow {}^{12}B + 2p$$
$$^{12}B \rightarrow {}^{12}C + \tilde{\nu}_e + e^-$$
$$\gamma + {}^{14}N \rightarrow {}^{12}N + 2n$$
$$^{12}N \rightarrow {}^{12}C + \nu_e + e^+$$
$$\gamma + {}^{13}C \rightarrow {}^{12}B + p$$
$$^{12}B \rightarrow {}^{12}C + \tilde{\nu}_e + e^-$$

The first two reactions are of interest, the third reaction provides the sole interfering signal. To use nitrogen as the marker in screening for explosives, the nitrogen signal must be separated from that of carbon.

In the image measurements performed [12.5] a $1 \, m^2$, $2 \, cm$ thick aluminum plate was mounted orthogonal to, and centered on, the electron beam. The plate served both as a support for the targets to be imaged and provided a background typical in kind and intensity to that which might be encountered in a realistic measurement. This plate was moved at a constant velocity of $\sim 6 \, cm/s$ transverse to the beam, which was incrementally scanned vertically in $2 \, cm$ steps. In the resulting pixels $\sim 85\%$ of the beam was contained within a $\sim 2 \, cm$ diameter circle for the direct electron beam, and a ~3 cm circle when the tantalum radiator was in place. The resulting saw tooth pattern produced images of 180 pixels in ~2.5 min.

A 5 cm thick box, filled with liquid scintillator and optically coupled to a fast photomultiplier tube, served as a detector of secondary radiation.

A typical time spectrum, and its deconvolution, of the response from a nitrogenous material is shown in Fig. 12.4, where the explosive simulating test object was melamine ($C_3H_5N_6$). In each of the four panels the horizontal scale corresponds to 100 ms subdivided in bins of 0.4 ms. The vertical scale gives the logarithm of the number of counts. There are 593 294 total events distributed with 74 843 in the first bin and 434 in the last. The spectrum was taken with 1 681 microtron pulses.

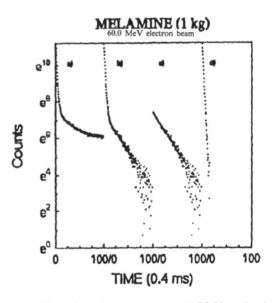

Figure 12.4. Melamine time spectrum. 50 MeV excitation. (a) raw spectrum, (b) without constant background, (c) nitrogen signal, (d) short-time background (from [12.4]).

The four panels of the figure show:

a) The raw spectrum, with a persistent background, equal in each bin, and a high short-time peak, probably originating from radiative neutron capture.

b) The same spectrum with the long-time background subtracted.

c) The isolated nitrogen signal, containing about 17% of the counts. Its slope corresponds to a half-life of 18.6 ms, which, when apportioned between ^{12}N and ^{12}B, gives 83% to the latter.

d) The short-time background with 1.3 ms half-life and containing about 63% of all events. The resultant nitrogen signal per pulse is around 50 counts above a negligible background.

An intensity image of ~125 g of plastic explosive SEMTEX attached to the plate and irradiated with 50 MeV electrons is clearly discernible in Fig. 12.5. This explosive contains carbon and nitrogen in about equal amounts.

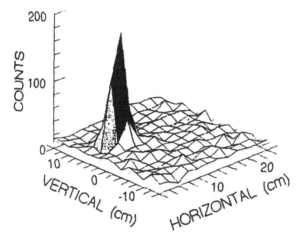

Figure 12.5. Intensity image of a cylinder of the plastic explosive SEMTEX irradiated directly by 50 MeV electrons (from [12.5]).

Certainly, it is only the first step in solving the problem of imaging the spatial distribution of nitrogeneous materials. Much remains to be accomplished before the nitrogen camera becomes a practical device.

12.3 Production of Radionuclides

The requirements for various good quality radionuclides are considerably high at present. The main production sources of radionuclides — nuclear reactors and cyclotrons — are not able to produce all types of radionuclides, therefore electron accelerators such as linacs and microtrons, that can produce radionuclides using photonuclear reactions, are suitable complements.

Although one can calculate the yields of photonuclear reactions taking into account the known values of the cross sections, the most reliable way is to measure them directly.

The yields for a number of radionuclides produced by bremsstrahlung irradiation were measured by different groups. Oka *et al.* [12.6] measured the yields of radioactivities induced by (γ, n) reactions with 20 MeV bremsstrahlung from a linac on 37 elements. The yields values obtained are plotted in Fig. 12.6, marked with the residual nuclides, against the atomic number of the nuclides.

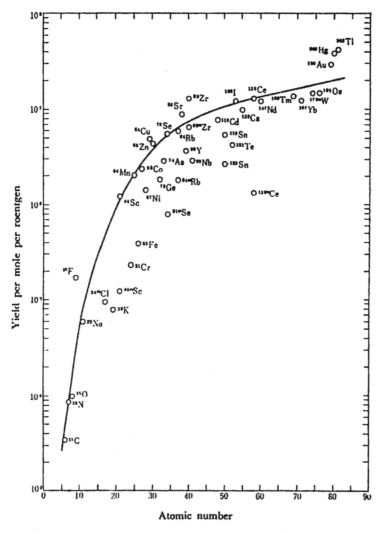

Figure 12.6. The yields of radionuclides from (γ, n) reactions as a function of atomic number with 20 MeV bremsstrahlung (from [12.6]).

The values generally increase regularly with atomic number

from 10^3 through 10^7 order of magnitude. This behaviour is analogous to that of the yields of photoneutrons which obtained as a function of Z values. This can also be well approximated from Levinger-Bethe's theory, one of the fundamental predictions of the theories arising from the application and extension of the sums rules for dipole photon-absorption to absorption by nuclei.

Analogous measurements were made at the Czech microtron by Řanda *et al.* [12.7] with 19 MeV bremsstrahlung. Detailed measurements were also performed in Dubna at the microtron for practically all natural nuclides at different electron energies [12.8]. To test the applied method the authors first compared the measurements of the yield of the photonuclear reaction $^{35}Cl(\gamma, n)^{34m}Cl$ with known data [12.6, 12.9, 12.10]. For this the samples of high-purity $SnCl_2 \cdot 2H_2O$ were irradiated by bremsstrahlung of electrons with different energies and the yield Y of ^{34m}Cl was calculated as

$$Y = \frac{A_0}{nI[1 - \exp(-\lambda t)]}, \qquad (12.1)$$

where Y was measured in decays per mole and per roentgen, I is the intensity of γ-flux in roentgen per min, t is time of irradiation in min, λ is constant of radioactive decay of ^{34m}Cl $(0.0214 \text{ min}^{-1})$. The results are given in Table 12.1.

Table 12.1. Yields of the photonuclear reaction $^{35}Cl(\gamma, n)^{34m}Cl$, measured in different experiments.

Electron energy, MeV	Yield, decay/mol·roent	Reference
14	$(3.05 \pm 0.5) \cdot 10^2$	
18	$(4.3 \pm 0.2) \cdot 10^4$	
18.5	$(7.0 \pm 0.2) \cdot 10^4$	[12.9]
19	$(8.4 \pm 0.1) \cdot 10^4$	
20	$9.1 \cdot 10^4$	[12.6]
21	$(1.2 \pm 0.05) \cdot 10^5$	
25	$(1.6 \pm 0.02) \cdot 10^5$	[12.9]
30	$2.2 \cdot 10^5$	[12.10]

It is clear that there is rather a good coincidence of different measurements.

Using the same method the yields of photonuclear reactions for 37 elements were measured and the results are presented in Table 12.2. Besides the light elements as C, N, O, and F, many radionuclides can be produced by a microtron, by means of photonuclear reactions, with high economic effect. In particular, the production of some radionuclides for medical preparations in nuclear medicine are very important. High activities up to giga-becquruels of ^{123}I, ^{11}In, ^{18}F and many other radionuclides can be prepared by a microtron, as was considered in Chapter 11.

Table 12.2. Experimental yields of photonuclear reactions (irradiation time 10 min)

Reaction	Half-life	E_γ, keV	Yield, dec/mol·r 14 MeV	18 MeV	20 MeV
^{26}Mg$(\gamma,p)^{25}$Na	60 s	975.2	4.09+E3	1.22+E4	2.01+E5
^{35}Mg$(\gamma,n)^{34m}$Cl	32 m	146.5	4.32+E3	9.56+E4	1.43+E5
^{48}Ti$(\gamma,p)^{47}$Sc	82 h	159.4	1.78+E3	2.27+E4	1.41+E5
^{50}Cr$(\gamma,n)^{49}$Cr	42 m	152.6	9.03+E3	1.75+E6	4.42+E6
^{57}Cr$(\gamma,p)^{56}$Mn	2.57 h	846.6	1.76+E4	1.83+E5	2.38+E5
		1810.7	1.87+E4	2.08+E5	1.05+E7
^{58}Cr$(\gamma,n)^{57}$Ni	36 h	127.0	2.97+E3	5.25+E4	2.81+E5
		1377.0	3.95+E3	5.88+E4	2.93+E5
^{59}Co$(\gamma,n)^{58}$Co	1711 h	810.6	3.9+E2	2.94+E3	2.98+E3
^{64}Zn$(\gamma,n)^{63}$Zn	38.4 h	669.6	9.49+E4	1.10+E6	1.91+E7
		961.9	1.01+E5	1.20+E6	2.15+E7
		1411.9	1.11+E5	2.0+E6	2.32+E7
^{70}Ge$(\gamma,n)^{69}$Ge	39 h	574.0	5.72+E4	2.74+E5	3.45+E5
		872.0	5.72+E4	3.21+E5	2.86+E5
		1106.5	5.17+E4	2.99+E5	3.12+E5
^{76}Ge$(\gamma,n)^{75}$Ge	82.2 m	199.2	2.32+E4	4.36+E5	3.25+E6
		264.8	2.47+E4	4.17+E5	3.28+E6
^{75m}Ge	46 s	139.8	4.19+E4	1.42+E5	2.65+E6
^{86}Sr$(\gamma,n)^{85m}$Sr	68 m	151.3	5.84+E5	7.10+E6	1.56+E7
		231.7	5.70+E5	6.94+E6	1.46+E7
^{88}Sr$(\gamma,n)^{87m}$Sr	2.8 h	388.4	2.26+E5	2.77+E6	5.28+E6
^{90}Zr$(\gamma,n)^{89}$Zr	78.4 h	909.1	6.75+E3	1.32+E4	2.14+E5
89mZr	4.18 m	587.8	3.12+E6	2.22+E7	8.95+E7
		1508.0	3.98+E6	2.54+E7	1.02+E9
^{93}Zr$(\gamma,n)^{92m}$Zr	240 h	934.5	7.31+E3	4.91+E4	1.61+E5
^{100}Mo$(\gamma,n)^{99}$Zr	66 h	739.7	2.14+E4	8.08+E4	1.06+E5
^{106}Cd$(\gamma,n)^{105}$Cd	56 m	347.1	1.40+E6	9.04+E6	8.27+E7
		606.6	1.32+E6	7.00+E6	9.20+E7
		961.1	1.13+E6	1.35+E7	2.26+E8
		1301.8	2.04+E6	1.20+E7	1.32+E8
^{112}Cd$(\gamma,n)^{111m}$Cd	48 m	150.6	4.49+E5	3.48+E6	2.73+E7
		245.3	5.22+E5	3.59+E6	3.13+E7

1	2	3	4	5	6
$^{116}Cd(\gamma,n)^{115}Cd$	53.5 h	492.3	2.56+E5	7.77+E5	5.97+E6
		527.8	2.85+E5	6.77+E5	5.98+E6
$^{113}In(\gamma,n)^{112}In$	14.4 m	606.4	1.11+E6	7.01+E6	1.05+E8
		618.2	1.20+E7	5.81+E7	9.90+E7
^{112m}In	20.7 m	155.4	1.31+E5	7.02+E5	1.25+E6
$^{112}Sn(\gamma,n)^{111}Sn$	35 m	761.7	4.75+E6	4.67+E7	2.66+E8
		1152.5	7.84+E6	4.74+E7	7.87+E7
		1913.8	9.35+E6	3.68+E7	1.14+E8
$^{124}Sn(\gamma,n)^{123m}Sn$	39.5 m	159.7	3.32+E7	2.85+E8	3.94+E8
$^{123}Sb(\gamma,n)^{122}Sb$	2.68 d	564.1	4.42+E5	6.08+E5	6.99+E5
$^{121}Sb(\gamma,n)^{120m}Sb$	15.9 m	1171.2	7.11+E7	1.02+E8	1.18+E8
$^{130}Te(\gamma,n)^{129m}Te$	69.6 m	459.5	1.29+E7	3.89+E7	5.92+E7
		487.4	1.36+E7	4.16+E7	6.32+E7
$^{133}Cs(\gamma,n)^{132}Cs$	6.47 d	667.6	1.53+E5	6.08+E5	1.28+E6
$^{138}Ba(\gamma,n)^{137m}Ba$	2.55 m	661.6	7.01+E5	4.14+E6	6.85+E7
$^{140}Ce(\gamma,n)^{139}Ce$	137 d	165.85	1.52+E4	2.94+E4	9.47+E4
^{139m}Ce	54 s	754.4	1.26+E7	3.33+E7	1.64+E8
$^{142}Nd(\gamma,n)^{141}Nd$	60.3 s	756.5	2.30+E6	2.46+E7	9.07+E7
$^{150}Nd(\gamma,n)^{149}Nd$	1.73 h	144.3	4.25+E6	8.32+E6	3.57+E7
		211.3	5.96+E6	1.09+E7	4.18+E7
		270.3	6.06+E6	1.46+E7	3.97+E7
$^{144}Sm(\gamma,n)^{143m}Sm$	65 s	754.4	6.88+E7	4.35+E8	1.50+E9
$^{160}Gd(\gamma,n)^{159}Gd$	18.6 h	363.5	2.80+E6	5.34+E6	1.23+E7
$^{169}Tm(\gamma,n)^{168}Tm$	93 d	198.0	1.79+E4	3.28+E4	9.82+E4
		815.0	3.11+E4	4.92+E4	1.02+E5
$^{162}Dy(\gamma,p)^{161}Tb$	6.9 d	74.6	1.21+E5	3.72+E5	1.67+E6
$^{170}Yb(\gamma,n)^{169}Yb$	31 d	110.0	8.23+E6	2.46+E7	7.38+E7
		197.8	1.45+E6	4.98+E7	1.25+E8
$^{176}Yb(\gamma,n)^{175}Yb$	4.2 d	282.6	7.64+E6	5.82+E7	1.79+E8
		396.1	1.07+E7	7.97+E7	3.17+E8
$^{180}Hf(\gamma,n)^{179m}Hf$	18.6 s	215.5	5.89+E6	5.31+E7	1.47+E8
$^{186}W(\gamma,n)^{185m}Hf$	1.6 m	131.4	3.50+E6	1.48+E7	1.81+E7
		188.2	4.33+E6	1.83+E7	2.66+E7
$^{192}Os(\gamma,n)^{191}Os$	15 d	129.5	9.23+E5	3.80+E6	1.27+E7
$^{191}Ir(\gamma,n)^{190}Ir$	11 d	518.4	4.57+E6	1.44+E7	7.27+E7
$^{193}Ir(\gamma,n)^{192}Ir$	74 d	316.5	6.57+E6	1.46+E7	6.35+E7
		468.1	7.36+E6	1.43+E7	5.89+E7
$^{192}Pt(\gamma,n)^{191}Ir$	72 h	538.9	5.78+E5	1.74+E6	5.23+E7
$^{197}Au(\gamma,n)^{196}Au$	6.2 d	355.7	5.34+E6	1.41+E7	6.29+E7
$^{198}Hg(\gamma,n)^{197m}Hg$	23.8 h	133.9	4.23+E4	3.93+E5	1.86+E5
		278.9	7.22+E5	2.83+E6	1.61+E6
$^{199}Hg(\gamma,p)^{198}Au$	2.7 d	411.8	3.27+E3	9.04+E3	2.61+E4
$^{200}Hg(\gamma,n)^{199m}Hg$	42.6 m	158.4	1.26+E6	5.07+E6	5.69+E6
		374.1	2.60+E6	9.94+E6	8.82+E6
$^{204}Pb(\gamma,n)^{203}Pb$	52.2 h	279.2	6.16+E6	1.68+E7	7.35+E7
$^{238}U(\gamma,n)^{237}U$	6.75 d	59.54	9.38+E6	5.45+E7	1.26+E8

12.4 Radiation Processing

High current accelerators in the energy range of 5 to 10 MeV, for use in industrial irradiation applications, have found a growing interest during the last years. The beam power of interest at 10 MeV electron energy is normally in the 20 to 50 kW range. At 5 MeV the power requirements are generally one magnitude larger, as electrons at this energy are often used for bremsstrahlung generation with a rather low electron to photon conversion efficiency ($\sim 5\%$). The accelerator type used is mostly the linear accelerator at 10 MeV; at 5 MeV high power DC-type accelerators are commonly used.

As it was analysed by Karlsson [12.11], the race-track microtron equipped with a small standing wave linac might offer a competitive alternative to the linac in terms of size, beam quality, stability, power conversion efficiency and cost.

The common factor in radiation processing is the use of large doses of radiation to effect products in a specific biological, chemical or physical way.

Polymerisation

This is a process whereby relatively simple molecules, the monomers, link up together to form high molecular weight units with desirable physical properties — such as polyethylene. Radiation breaks chemical bonds, makes free radicals and allows the bonding of the monomers.

Radiation polymerisation can be achieved for products which cannot be synthesised by the classical methods which initiate polymerisation such as heat, pressure, photochemical or electrochemical processes. Polymers obtained in that way exhibit high resistance to temperature and chemicals. A related technology is graft polymerisation which consists of polymer chains built from two different monomers. This allows the synthesis of materials with complementary properties such as highly adherent films (polyethylene grafted with acrylic acid) or fibres and textiles with features such as wrinkle, shrinkage or heat resistance, dye stability, washability or antistatic properties.

Cross-linking

Irradiation of fully polymerised material gives rise to two competing processes: cross-linking and degradation. Linear molecular chains can be further linked with each other through additional branching — leading to spatial cross-linking and generating in effect a polymer with a three dimensional lattice with properties totally different from those of the unidimensional compound.

The cross-linking process is used to produce heat-shrinkable films for packaging, and to make components for the electric and gas industries. A particularly important application is the treatment of wire and cable insulation to give them a high mechanical resistance and the capacity to withstand thermal stresses (overloads and short circuits). Through radiation processing one can reduce the insulation thickness and therefore the wiring size for complex control systems and telecommunication facilities (telephone exchanges). Cross-linking of elastomers is called vulcanisation and is used in the tyre industry to treat rubber. Radiation cross-linking is the most commonly used industrial radiation process.

Curing of coatings

The lifetime, and other desirable properties (surface hardness, heat resistance, etc.) of many products are achieved through the proper processing of their coating, called curing, which consists of a particular form of copolymerisation or cross-linking. These chemical reactions can be achieved by means of high energy electrons (up to 300 kV). Radiation curing of paint and varnishes may achieve substantial energy savings compared with conventional thermal processing.

Degradation

Some polymers instead of cross-linking are degraded upon high energy irradiation. This is interesting for the recovery of plastic wastes and their transformation into useful products. For instance, teflon wastes can be degraded into lower molecular weight compounds and recycled for the production of lubricant oil.

Degradation processes are also expected to find large application in the agricultural product and food industry (e.g. processing of molasses into alcohol).

Sterilisation

It has been know since the early days following the discovery of X-rays that micro-organisms die under the action of radiation. There is a large need to sterilise medical material, such as coats, dressing materials, syringes, catheters, transfusion and injection sets. The traditional sterilisation method is steam at 150°C but it is not tolerated by many of the low cost disposable products made from plastics. The chemical sterilisation method relying on ethylene oxide (ETO) also has drawbacks as ETO is not only a poison for bacteria but is also hazardous and may cause chronic diseases to health personnel. These considerations explain the high interest in radiation sterilisation. However, the diffusion of the process requires conforming to strict standards. Care must for instance be taken to avoid the appearance of undesirable radiation degradation products which could be highly toxic. This is for instance the case in PVC which may produce hydrochloric acid after radiation degradation.

Food Preservation

The idea for using high energy ionising radiation to preserve food has been around for over 50 years. The economic importance of this application can easily be appreciated if one considers that in many countries some 20% of the food production is lost because of improper storage and lack of appropriate preservation measures. Radiation can avoid the sprouting of products such as potatoes or onions. Radiation is also able to destroy insects, parasites and pathogens and to prolong the storage duration of poultry, fish or fresh fruits while preserving their nutritional value. As for sterilisation, food preservation by radiation would avoid the use of chemical disinfectants which are not only toxic but even carcinogenic.

In spite of numerous tests and studies, this application remains rather limited. The primary reason is the extreme caution from regulating agencies when dealing with any radiation related processes, the more so if it concerns the area of human nutrition. It should also be recognised that the energy and dose must be carefully adjusted so as to avoid possible unpleasant side effects resulting from unwanted radio-chemical reactions which could modify the taste or the aspect of the products.

Pollution Control

High energy radiation produced by accelerators can contribute in various ways to the reduction of environmental pollution and the associated health hazards.

- The treatment of waste water and sewage sludge is a major problem in all countries. Radiation can destroy toxic bacteria and permit the recycling of water which becomes in short supply in many places, and the safe use as a fertiliser of sewage sludge and excreta from farm animals which still have a high nutrient content.

- Electron beams can ionise the soot particles contained in the gaseous effluents from power stations, chemical or waste incineration plants. They can then be removed by electrostatic precipitators. The same process has been shown to be capable of trapping SO_2 and NO_x through conversion into complex ammonium salts, if a small amount of ammonia is added beforehand.

Other applications

Some other applications are

• Hardening of metal surfaces where an intense electron beam heats a metal surface for a very short period. The metal surface is then rapidly cooled by heat transfer into the deeper parts of the material.

• Electrical characteristics of semiconductors, especially switching characteristics and forward resistance in thyristors, can be affected by radiation.

• Ancient objects of organic materials can be preserved if they are irradiated so that the microorganisms that destroy them are killed.

12.4.1 Degradation of Phenolic Substances Under Electron Irradiation

As an example of the efficiency of the use of electron accelerators in waste water purification we consider below experiments which were performed at the ENEA Research Center at Frascati (Italy).

12.4.1.1 Irradiation of Phenols

Polyphenols are well known to act as inhibitors of the microbial degradation of olive-crushing waste waters during lagooning, which is commonly used as disposal technique. A rough estimate the production of the olive oil industry in Italy shows that it is in the range of $(3 \div 5) \times 10^6 \, m^3/yr$. Long-lasting disposal of these huge volumes of effluents, which are sources of olfactory pollution and infection risks, is required. On the other hand, the options so far investigated as to the exploitation or disposal of these waste waters (including irrigation, lagooning, thermal treatment, concentration, and so on) look controversial. Specific phenol-removal techniques (including ultrafiltration and reversed osmosis) do not work satisfactorily in the mill effluents.

In the framework of an ENEA project aimed at developing electron accelerators for industrial applications a set of experiments were undertaken concerning the destruction of polyphenols in order to accelerate the degradation by specific bacterial cultures of the olive-mill waste waters. The electron irradiation of phenols as a model compound was performed.

The phenol solutions were irradiated in the aforementioned 5 MeV linear accelerator with a dose of up to 20 Mrad. A photometric procedure based on the colour reaction of phenols with 4-amioantipyrine was adopted as an analytical tool. The degradation curves for phenol concentrations of $(0.25 \div 3.5)$ g/l to doses of $(0.5 \div 10)$ Mrad are shown in Fig. 12.7.

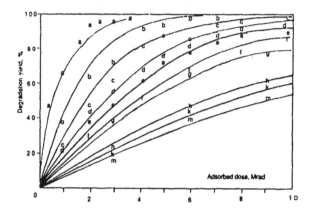

Figure 12.7. Phenol degradation under electron irradiation; legenda, phenol concentration in g/l: (a) 0.25, (b) 0.5, (c) 0.75, (d) 1.0, (e) 1.25, (f) 1.75. (g) 2.25, (h) 2.5, (k) 3.0, (m) 3.5.

The following exponential relationship holds for the degradation of phenolic molecules:

$$Y = 100[1 - \exp(D/D_0)], \qquad (12.2)$$

where Y is the degradation yield in %, D the adsorbed dose in Mrad and D_0 a parameter depending on the initial phenol concentration C_0 (see Fig. 12.8) according to the relationship $D_0 = KC_0$ with $K = 3.3$ Mrad/gl.

From these data, an empirical factor E accounting for the decomposition energy of the phenolic molecules has

$$E = \left[\frac{D_0}{C_0}\right]^{-1}\left[10^3 \frac{M}{N_A}\right] = 48 \text{ eV/molecule}. \qquad (12.3)$$

where M is the phenol molecular weight in g and N_A is the Avogadro number.

12.4.1.2   Irradiation of Pesticides

Atrazine, parathion and hexachlorobenzene are commonly found in soils and groundwaters due to their widespread use in agriculture as herbicide, insecticide and fungicide, respectively. They represent a serious contamination problem, especially where drinking water is concerned, and the current water clarification treatments do not seem to be as efficient as desirable.

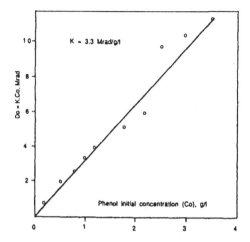

Figure 12.8. D_0 vs phenol initial concentration under electron irradiation.

The pesticide samples were irradiated by 5 MeV electrons to a dose of $(0.5 \div 8)$ Mrad. Gas-chromatographic procedures with an electron-capture detector for parathion and hexachlorobenzene and a thermionic $(n - p)$-detector for atrazine were used. The degradation curves for the pesticides investigated in [12.12] are shown in Fig. 12.9. Altrazine and parathion in aqueous solution were found to be completely decomposed.

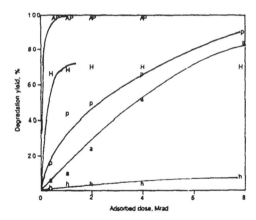

Figure 12.9. Degradation curves of atrazine, parathion and hexachlorobenzene under electron irradiation; atrazine: $a =$ solid in air $(20 \mu$ g$)$, $A =$ solution $(1.9 \mu$ g/l$)$, hexachlorobenzene: $h =$ solid in air $(24 \mu$g$)$, $H =$ solution $(5.5 \mu$ g/l$)$, parathion: $p =$ solid in air $(24 \mu$ g$)$, $P =$ solution $(1.4 \mu$ g/l$)$.

References

Chapter 1

1.1. Veksler V.I., *Dokl. Akad. Nauk USSR* **43** (1944) 329
1.2. Alvarez L.W. (1939). In: H.F.Kaiser, *J. Franklin Inst.* **259** (1954) 89
1.3. Itoh J. and Kobayashi D., Kyoto meeting (1945). In: Sci. Papers Osaka Univ. 12 (1949); also in Kumagaya H., Particle Accelerator. Kyoritsu Publishing Co. Tokyo, 1975
1.4. Schwinger J.S., Harward Lectures (1945). In: Schiff L.I. *Rev. Sci. Instr.* **17** (1946) 9
1.5. Kapitza S.P., Bykov V.P., Melekhin V.N., *Sov. Phys. JETP* **41** (1960) 997
1.6. Sedláček M., Stockholm Conf. (1955). In: G.Borelins and E.Rudberg, *Ark. Fys.* **11** (1956) 129
1.7. Wernholm O., *Ark. Fys.* **26** (1964) 527
1.8. Sigiyama K., Takafuji A., Kuroda K. *et al. IEEE Trans. Nucl. Sci.* **44** (1997) 1673
1.9. Melekhin V.N., *Soviet Phys., JETP* **15** (1962) 433
1.10. Kolomensky A.A., *J. Techn. Phys.* **30** (1960) 1347
1.11. Henderson C., Heymann F.F., and Jenningd R.S. *Proc. Roy. Soc.* **B 66** (1953) 654
1.12. Sells V., Froelich H, and Brannen E., *J. Appl. Phys.* **36** (1965) 3264
1.13. Bizzarri U., Messina G., Mola A., Picardi L. and Vignati A., *Nucl. Instr. Meth.* **B 29** (1987) 573

Chapter 2

2.1. Belov A.G. Proc. Workshop on Application of Microtrons in Nuclear Physics, Plovdiv, 22-24 September 1992, D15-93-80, JINR, 1993, p.12
2.2. Rowe E.M. and Mills F.E. *Particle Accelerators* **4** (1973) 211
2.3. Bizzari U., Messina G., Mola A. *et al., Nucl. Instr. Meth.* **B 29** (1987) 573
2.4. Aleshin V.M., Doronin A.V., Lebedev V.I. *Instr. Exper. Techn.* **33** (1990) 525
2.5. Melekhin V.N. *Instr. Exp. Techn.* **33** (1990) 290

2.6. Kazakevich G.M., Marusov V.N., Silvestrov G.I. Proc. II Asian Symp. on Free Electron Lasers, Novosibirsk, 1995, p. 257

2.7. Belov V.P., Karasjuk V.N., Kazakevich G.M. *et al.* Proc. II Asian Symp. on Free Electron Lasers, Novosibirsk, 1995, p. 253

2.8. Wainstein L.A., High Power Electronics, **3** (1964) 216, Moscow, Nauka

2.9. Zhulinskii S.F., Luk'yanenko E.A., and Mirzoyan A.R. *Instr. Exper. Techn.* **14** (1971) 353

2.10. Alekseev I.V., Vladimirov N.V., Gorbachev V.P., Solovyev A.V. and Stepanchuk V.P. Proc VIII Meet. on Acc., v.ll, p.41, JINR (Dubna) 1983

2.11. Reich H. and Löns K. *Nucl. Instr. Meth.* **31** (1964) 221

2.12. Shaw E.D., Mills A.P., Chichester R.J. *et al.*, *Nucl. Instr. Meth.* **B 56/57** (1991) 568

2.13. Gridnev V.I., Rozum E.I., Slupskii A.M., and Vorob'ev S.A. *Instr. Exp. Techn.* **30** (1987) 16

2.14. Reich H. *Z.Naturforsch.* **13a** (1958) 1003

2.15. Svensson H., Jonsson L., Larsson L.-G. *et al.*, *Acta Radiologica Therapy Physics Biology* **16** (1977) 145

2.16. Belovintsev K.A., Belyak A.Ya., Grornov A.M. *et al.*, *Atomic Energy* **14** (1963) 359

2.17. Vorob'ev A.A., Chuchalin I.P., Vlasov A.G. *et al.*, Synchotron TPI for 1.5 GeV. M.: Atomizdat, 1968

2.18. Rowe E.M., Green M.A., Trseciak W.S., and Winter W.R. Proc. IX Int. Conf. High Energy Acc., 689, SLAAC (1974)

2.19. Green M.A., Rowe E.M., Trzeciak W.S., and Winter W.R. *IEEE Trans. Nucl. Sci.* **NS-28** (1981) 2074

2.20. Wernholm O., *Ark. Fys.* **26** (1964) 527

2.21. Ramamurthi S.S. and Singh G. *Nucl. Intsr. Meth.* **A 359** (1995) 15

2.22. Botman J.I.M., van Genderen W,, HagedoomH.L. *et al.*, Proc. 1st Eur. Part. Acc. Conf., Rome, **1** (1988) 453

2.23. Ziyakaev R.G., Onisko A.D., Chuchalin I.P. *et al.*, Electron Accelerators, Proc.VII Interhighschool Conf. Electr. Accel., vyp.l, Atomizdat, 1970, Moscow, p.20

2.24. Wolski W. Report No. 69-12, KTH, Stockholm, 1969

2.25. Rosander S. and Wernholm 0. ALA-1994-104, KTH, Stockholm, 1994

2.26. Melekhin V.N. *Nucl. Instr. Meth.* **A 337** (1993) 29

2.27. Seryapin V.G., Patrenin V.A., and M.Limasov Yu.M., *Instr. Exp. Techn.* **14** (1971) 350

2.28. Kuznetsov G.I. *Instr. Exp. Techn.* **40** (1997) 424

2.29. Min-quan Qian, Mao-rong Yang, Qing Pan *et al.* *Nucl. Instr. Meth.* **A 358** (1995) 280

2.30. Srinivasan-Rao T., Fischer J. and Tsang T. *J. Appl. Phys.* **69** (1991) 3291

2.31. Asakalna M. *et al.*, *Nucl. Instr. Meth.* **A 331** (1993) 302

2.32. Takafuji A., Sugiyama K., Kuroca K. *et al.* *IEEE Nucl.Sci.* **44** (1997) 1677

2.33. Gal E.G., Eponeshnikov V.N, and Lashuk N.A. *Instr. Exp. Techn.* **28** (1985) 18

2.34. Kanter B.Z. *Izv. Vyssh. Uchebn. Zaved., Fizika* No. 3 (1960) 138

2.35. Esina Z.N., Kuznetsov V.M., and Lashuk N.A., *Pryb. Tech. Eksp.* No.3 (1981) 24

2.36. Derbenev Ya.S. and Kondratenko A.M. *Dokl. Akad. Nauk SSSR* **223** (1975) 830

2.37. Luk'yanenko E.A. and Melekhin V.N. *Instr. Exp. Techn.* **19** (1976) 981

2.38. Melekhin V.N., Luk'yanenko E.A. *High Power Electronics* **5** (1968) 257

2.39. Luk'yanenko E.A., Melekhin V.N. *J. Techn. Phys.* **40** (1970) 414

2.40. Luganskii L.B., Candidate's Dissertation, Institute for Physical Problems, Academy of Sciences of the USSR, Moscow, 1970

2.41. Melekhin V.N. *High Power Electronics* **5**, Nauka, Moscow (1968), p. 228.

2.42. Kapitza S.P., Melekhin V.N., Zakirov B.S. *et al. Instr. Exp. Tech.* No. 1 (1969) 13

2.43. Aleshin V.M. and Doronin A.V. *Instr. Exp. Techn.* **34** (1991) 26

2.44. Zakharov M.A., Lugansky L.B., Melekhin V.N. *et al., J. Tech. Phys.* **47** (1977) 593

2.45. Semenov V.K. *J. Techn. Phys.* **37** (1992) 1162

2.46. Kapitza S.P., Bykov V.P., Melekhin V.N. *JETP* **41** (1961) 368

2.47. Belovintsev K.A., Levonyan S.V., Serov A.V. *J. Techn. Phys.* **51** (1981) 752

2.48. Belovintsev K.A., Serov A.V. *Sov. Phys. - Lebedev Institute Reports* **9** (1983) 18

2.49. Rosander S. *Nucl. Instr. Meth.* **56** (1967) 154

2.50. Stavisskii Yu.Ya., Proc. II All-Union Conf. Acc. Charged Part., v.l, M.: Nauka, 1972, p.78

2.51. Rosander S. *Nucl. Instr. Meth.* A **236** (1985) 222

2.52. Zav'yalov V.V. and Semenov V.K. *Instr. Exp. Techn.* **30** (1987) 1301

2.53. Rodionov F.V., Stepanchuk V.P., *J. Techn. Phys.* **41** (1971) 999

2.54. Abramov A.I., Slobodyanyuk V.A., Kapitza S.P., Proc. VIII All-Union Conf. Accele., v. II (1983) 38

2.55. Kapitza P.L., Filimonov S.I., High-Power Electronics, N 6, M.: Nauka, 1969, p.7

2.56. Alekseev I.V., Vladimirov N.V., Gorbachev V.P., Soloviev A.V., Stepanchuk V.P. Proc. VIII All-Union Conf. of Charge Particle Accel., Dubna, 1983, v.II, p.41

2.57. Polyakov V.I., Rodionov F.V., Stepanchuk V.P. *Sov.Phys.-Tech.Phys.* **16** (1972) 1311

2.58. Balaev A.Yu., Vishnevskiii A.A., Dudkin V.N. *et al., Instr. Exp. Techn.* **33** (1990) 1419

2.59. Kashtanov V.V., Saprygin A.V., Stepanchuk V.P., and Balaev A.Yu., *Instr. Exp. Techn.,* **35** (1992) 771

2.60. Brezhnev O.N., kazakevich G.M., Ponomarchuk V.A., Filimonov E.M.,
       Instr. Exp. Techn., **18** (1975) 338

Chapter 3

3.1.   Schiff L.I. *Rev. Sci. Instr.* **17** (1946) 6
3.2.   Brannen E. and Froelich H.R. *J. Appl. Phys.* **32** (1961) 1179
3.3.   Knapp E.A., Knapp B.C., Potter J.M. *Rev. Sci. Instr.* **39** (1968) 979
3.4.   Froelich H.R., Tompson A.S., Edmonds D.S. and Manca J.J. *IEEE
       Trans. Nucl. Sci.* **NS-20** (1973) 260
3.5.   Babić H. and Sedlaček M. *Nucl. Instr. Meth.* **56** (1967) 170
3.6.   Rosander S., Sedlaěk M., Wernholm 0., Babić H. *Nucl. Instr. Meth.* **204**
       (1982) 1
3.7.   Eriksson M. *Nucl. Instr. Meth.* **261** (1987) 39
3.8.   Belovintsev K.A., Karev A.I., Kurakin V.G. *Nucl. Instr. Meth.* **261**
       (1987) 26
3.9.   Rand R.E. 'Recirculating Electron Accelerators'. Harwood Academic
       Publishers, 1984
3.10. Rosander S. *Nucl. Instr. Meth.* **177** (1980) 411
3.11. Picardi L. Report in Indo-Soviet Seminar on Microtrons, CAT, Indore
       1992, India
3.12. Karlsson M., ISRN KTH/ALA/R-93/1-SE (1993)
3.13. Picardi L., Giubileo G., Raimondi P., Ronsivalle C., Vignati A. Proc.
       2nd Eur. Part. Ace. Conf. 1990, Nice, v.1 (1990) p.443
3.14. Herminghaus H.A., Feder A., Kaiser K.H. *et al. Nucl. Instr. Meth.* **138**
       (1976) 1
3.15. Botman J.I.M., Xi B., Timmermans C.J., Hagedoorn H.L. *Nucl. Instr.
       Meth.* **B 68** (1992) 101
3.16. Delhez J.L., The Azimutally Varying Field Racetrack Microtron, Eind-
       hoven: Eindhoven University of Technology, ISBN 90-386-0343-6 (1994)
3.17. Axel P., Hanson A.0., Hailan J.R. *et al. IEEE Trans. Nucl. Sci.* **NS-22**
       (1975) 1176
3.18. Herminghaus H. and Euteneuer H. *Nuck. Instr. Meth.* **163** (1979) 299
3.19. Alimov A.S., Chepurnov A.S., Chubarov O.V. *et al.* Preprint INP
       MSU-93-9/301, Moscow 1993
3.20. Wilson M. *et al.* 1988 Linear Acc. Conf. Proc., CEBAF-Report-89-001,
       p.255
3.21. Kolomensky A.A. *J. Techn. Phys. Lett.* **5** (1967) 204
3.22. Messina G. *et al.* EPAC 1988, 1447
3.23. Herminghaus H. Proc. EPAC92, Editions Frontieres, Gif sur Yvette,
       France, 1992, p.247
3.24. Kaiser K.H. Proc. of the Conf. on Future Possibilities for Electron
       Accelerators, 1979. Charlottesville, VA
3.25. Jackson H.E. *et al.*, Argonne Publication ANL-82-83 (1982)
3.26. Belovintsev K.A. *et al.*, Preprint No.88, Lebedev Physical Institute,
       Moscow 1984

3.27. Herminghaus H. *Nucl. Instr. Meth.* **A 305** (1991) 1

3.28. Trower W.P., Karev A.I., Melekhin V.N. *et al. Nucl. Instr. Meth.* **B 99** (1995) 736

3.29. Shvedunov V.I., Karev A.I., Melekhin V.N. *et al.* Proc. 1995 Part. Acc. Conf., IPS II (1995) 807

3.30. Shelaev I.A. JINR Communications P9-91-31, Dubna 1991

Chapter 4

4.1. Kim K.-J. and Sessler A. *Science* **250** (1990) 88

4.2. Weizsäcker C.F.von *Z.Phys.* **88** (1934) 612

4.3. Williams E.J. *Kgl. Dansk. Vid. Selsk.* No.4 (1935) 13

4.4. Ellias L.R., Fairbank W.M., Madey J.M.J. *et al., Phys. Rev. Lett.* **36** (1976) 717

4.5. Deacon D.A.G., Ellias L.R., Madey J.M.J. *et al., Phys. Rev. Lett.* **38** (1977) 892

4.6. Dattoli G. and Torre A., CERN 89-03 (1989)

4.7. Shaw E.D., Chichester R.J. and Chen S.C. *Nucl. Instr. Meth.* **A 20** (1986) 44

4.8. Kapitza S.P., Bogomolov G.D., Wainshtein L.A., Zavialov V.V. *Nucl. Instr. Meth.* **A 259** (1987) 285

4.9. Ciocci F., Doria A., Gallerano G.P. *et al., Phys. Rev. Lett.* **66** (1991) 699

4.10. Ciocci F., Bartolini R., Doria A. *et al., Phys. Rev. Lett.* **70** (1993) 928

4.11. Shaw E.D., Chichester R.J. *Nucl. Instr. Meth.* **A 318** (1992) 47

4.12. Akberdin R.R., Kazakevich G.M., Kulipanov N.A. *et al. Nucl. Instr. Meth.* **A 405** (1998) 195

4.13. Halpbach K. *Nucl. Instr. Meth.* **A 246** (1986) 77

4.14. Vinokurov N.A. and Skrinsky A.N., INP preprint 77-59 (Novosibirsk, 1977)

4.15. Kapitza S.P. and Semenov V.K. *Nucl. Instr. Meth.* **A 308** (1991) 97

4.16. Kapitza S.P. *Nucl. Instr. Meth.* **A 261** (1987) 43

4.17. Wainstein L.A., Kleev A.I., Solntzev V.A. *Nucl. Instr. Meth.* **A 308** (1991) 94

4.18. Kapitza S.P., Kleev A.I. *Radiotechnica i Electronica [Soviet Journal of Communications Technology and Electronics]* **36** (1991) 2379

4.19. Vinokurov N.A., Gavrilov N.G., Gornikel E.I. *et al., Nucl. Instr. Meth.* **A 359** (1995) 41

4.20. Brau C.A., Boyd T.J., Cooper R.K., and Swenson D.A. High Efficiency Free-Electron Laser Systems, Proc. Int. Conf. on Lasers'79 (December 17-21, 1979)

4.21. Gavrilov N.V., Oreshkov A.D,et al., Pinayev I.V. *et al., Nucl. Instr. Meth.* **A 359** (1995) 44

Chapter 5

5.1. Tamm I.E. and Frank I.M. *Doklady Akad. Nauk SSSR* **14** (1937) 107
5.2. Ginzburg V.L. and Frank I.M. *ZhETF* **16** (1946) 15
5.3. Garibyan G.M. *ZhETF* **37** (1959) 527; **39** (1960) 332
5.4. Barsukov K.A. *ZhETF* **37** (1959) 1106
5.5. Cherry M.L., Hartman G., Muller D., and Prince T.A. *Phys. Rev.* **D 10** (1974) 3594
5.6. Fainberg Ya.B. and Thiznjak N.A. *ZhETF* **32** (1957) 883
5.7. Ter-Mikaeljan M.L. *Influence of Medium on Electromagnetic Processes by High Energy*, Erevan (Akad. Nauk Arm. SSR), 1969
5.8. Baryshevsky V.G. *Doklady Akad. Nauk BSSR* **15** (1971) 306
5.9. Baryshevsky V.G. and Feranchuk I.D. *ZhETF* **61** (1971) 944
5.10. Akhiezer A.I. and Berestetsky V.B. *Quantum Electrodynamics* (M., Nauka), 1969
5.11. Feranchuk I.D. and Ivashin A.V. *J. Phys.* **46** (1985) 1981
5.12. Andersen J.U., Augustyniak W.M. *Phys. Rev.* **B 3** (1971) 705
5.13. Terhune R.W and Pantell R.H. *Appl. Phys. Lett.* **30** (1977) 265
5.14. Kumakhov M.A. *Phys. Lett.* **57** (1976) 17; see also Pantell R.H., Swent R.L., Datz S. *et al. IEEE Nucl. Sci.* **NS-82** (1981) 1152, and Klein R.K., Kephart J.O., Pantell R.H. *et al. Phys. Rev.* **B 40** (1989) 4249
5.15. Genz H., Gräf H.-D., Hoffmann P.*et al. Appl. Phys. Lett.* **57** (1990) 2956

Chapter 6

6.1. Kondev Ph.D., Tonchev A.P., Khristov Kh.G., and Zhuchko V.E. *Nucl. Instr. Meth.* **B 71** (1992) 126
6.2. Lanzl L.H. and Hanson A.O. *Phys. Rev.* **89** (1953) 959
6.3. Tsovbun V.I., JINR 16-7104, 1973
6.4. Zhuchko V.E. and Tsipenyuk Yu.M., Atomnaya energiya (in Russian), v.27 (1975), p.66.
6.5. MacGregor M.H. *Nucleonics* **17** (1959) 104
6.6. Tsipenyuk Yu.M. *Atomnaya energiya (in Russian)* **27** (1969) 468
6.7. Lawson J.D. *Proc. Roy. Soc.* **A 63** (1950) 653
6.8. Rossi B. and Greisen K. *Rev. Mod. Phys.* **13** (1941) 240
6.9. Zhalsaraev B.Zh., Mescheryakov R.P., Yakovlev B.M. *et al. Prikladnaya yadernaya spektroskopiya* vyp.10, Moscow, Atomizdat, 1981, p.189.
6.10. Sincler R.H. and Day O.G. *Elementary Particles and Atomic Nuclei* **2** (1972) 981
6.11. Gayther D. and Goode P. *J. Nucl. Energy* **21** (1967)
6.12. Drobinin A.V., Leonard M. and Tsipenyuk Yu.M. *Atomnaya energia* **53** (1982) 398
6.13. Kimura I. *et al. Ann. Rep. Res. Reactor Inst.Kyoto Univ.* **3** (1970) 75
6.14. Thompson M., Taylor J., *Nucl.Phys.* **76** (1966) 377

6.15. Gozani T., Radiation Engineering in the Academic Curriculum, Vienna, IAEA, 1975, p.25

6.16. Kovalëv V.P. Vtorichnye izlucheniya uskoritelei elektronov, Moscow, Atomizdat, 1979

6.17. Kochenov A.S., Sadikov I.P., Stolypin V.S. Kurchatov Atomic Energy Institute. IAE-2461.1974; Kolmychkov N.V., Sidorkin S.F., Stavissky Yu.Ya. Institute of nuclear researches AS USSR, P-0174, Moscow, 1980

6.18. Piestrup M.A., Boyers D.G., Pincus C.I. *et al.*, *SPIE* **1552** (1991) 214

6.19. Piestrup M.A., Kephart J.O., Park H. *et al.*, *Phys. Rev.* **A 32** (1985) 917

6.20. Piestrup M.A., Moran M.J.,Boyers D.G. *et al.*, *Phys. Rev.* **A 43** (1991) 2387

6.21. Piestrup M.A., Boyers D.G., Pincus C.I. *et al.*, *Phys. Rev.* **A 43** (1991) 3653

6.22. Rule D.W., Fiorito R.B., Piestrup M.A. *et al.*, *SPIE* **1552** (1991) 240

6.23. Rule D., *Nucl. Istr. Meth.* **B 24/25** (1987) 901

6.24. Lumpkin A. *et al.*, *Nucl. Istr. Meth.* **A 296** (1990) 150

6.25. Rule D. *et al.*, *Nucl. Istr. Meth.* **A 296** (1990) 739

6.26. Genz H., Gräf H.-D., Hoffmann P. *et al.*, *Appl. Phys. Lett.* **57** (1990) 2956

6.27. Jolie J. and Bertschy M. *Nucl. Instr. Meth.* **B 95** (1995) 431

6.28. Bertschy M., Jolie J., Mondelaers W. *Nucl. Instr. Meth.* **B 95** (1995) 437

6.29. Bertschy M., Critin M., Jolie J. *et al.* *Nucl. Instr. Meth.* **B 103** (1995) 330

6.30. Bertschy M., Jolie J., Mondelaers W. *Appl. Phys.* **A 62** (1996) 437

6.31. Belovintsev K.A. and Cherenkov P.A., Proc. Int. Conf. on Accelerators, Dubna, 1963, p.1056

6.32. Belovintsev K.A. and Denisov F.P. *Atomic Energy* **16** (1964) 353

6.33. Eriksson M., Reistad D., Rosander S., ISSN 0348-7539 Report TRITA-EPP-79-05, 1979, Electron and Plasma Physics, Royal Inst. of Techn., Stokholm, Sweden

6.34. Kozhemyakin A.V., Petrov V.V., Yasnov G.I., Electron Accelerators, Proc. VII Interuniversity Conf. Electr. Acc., 1, Atomizdat 1970, p.23

6.35. Bernardini N. *et al.*, Centre d'Etudes Nucléaires de Saclay Report, June 1963

6.36. Sund R., Walton R., *Nucl. Instr. Meth.* **27** (1964) 109

6.37. Katz L., Lokan W., *Nucl. Instr. Meth.* **11** (1961) 7

6.38. Jupiter C.P., Hansen N.E., Shafer R.E., and Fultz S.C. *Phys. Rev.* **121** (1961) 866

6.39. Schultz P.J. and Lynn F.G. *Rev. Mod. Phys.* **60** (1988) 701

6.40. Coleman P.G., Sharma S.C., and Diana L.M. (eds.), Positron Annihilation (North-Holland, Amsterdam, 1982)

6.41. Dorikens-Vanraet L., Dorikens M., and Segers D., (eds.), Positron Annihilation (World Scientific, Singapore, 1989)

6.42. Mills A.P. Shaw Jr., E.D., Chichester R.J., and Zuckerman D.M. *Rev. Sci. Instr.* **60** (1989) 825

6.43. Mills A.P., Jr. *Appl. Phys.* **22** (1980) 273
6.44. Shaw E.D., Mills A.P., Jr., Chichester R.J. *et al. Nucl. Instr. Meth.* **B 56/57** (1991) 568

Chapter 7

7.1. Shchetinin A.M., Nikitin V.N. *Atomic Energy* **38** (1975) 97
7.2. Burmistenko Yu.N., Photonuclear Analysis of Substance Content (in Russian), Energoatomizdat, Moscow, 1986
7.3. Brovtsyn V.K., Samosyuk V.N., and Tsipenyuk Yu.M. *Atomic Energy* **32** (1972) 383
7.4. Tsipenyuk Yu.M., Firsov V.I. *J. Radioanal. Nucl. Chem.* **216** (1997) 47
7.5. Belov A.G. and Teterev Yu.G., JINR 18-84-8, Dubna 1984
7.6. Świderska-Kowalczyk M., Zóltowski T., Landeyro P.A *Nucleonika* **37** (1992) 17
7.7. Beckurts K.H. and Wirts K., *Neutron Physics*, Springer-Verlag, 1964

Chapter 8

8.1. Bergere R. *Features of the Giant E1 Resonance,* In: Photonuclear Reactions I, Springer-Verlag, 1977, p.1
8.2. Ratner B.S. *Elementary Particle and Atomic Nuclei* **12** (1981) 1492
8.3. Hayward E. *Photon Scattering in the Energy ange 5-30 MeV.* In: Photonuclear Reactions I, Springer-Verlag, 1977, p.340
8.4. Balabanov N.P., Belov A.G., Gangrsky Yu.P. *et al.* preprint JINR E15-93-37, Dubna, 1993
8.5. Kato T. *J. Radioanal. Chem.* **16** (1973) 307
8.6. Burmistenko Yu.N. Photonuclear Analysis of a Content of Substace. M.: Energoatomizdat, 1986
8.7. Caldwell J.T. and Dowby E.J. *J. Nucl. Sci. Eng.* **56** (1975) 179
8.8. Nikotin O.P. and Petrzhak K.A. *Atomic Energy* **20** (1966) 268
8.9. Gozani T., Bramblett R.L., Ginaven R.O., Runquist D.E. *Int. Conf. Photonucl. Reactions and Applications*, Asilomar, v.1 (1973) p.508

Chapter 9

9.1. Mezhiborskaya K.B. Photoneutron Method for Determining Beryllium (in Russian). Gosatomizdat, Moscow, 1961
9.2. Hevesy G. and Levi H. *Math.-Fys. Meddeleser* **14** (1936) 3
9.3. Seaborg G.T. and Livingood J.J. *J. Amer. Chem. Soc.* **60** (1938) 1784
9.4. Řanda Z. and Kreisinger F. *J. Radioanal. Chem.* **77** (1983) 279
9.5. Zykin L.M., Kolosov A.A. and Tsipenyuk Yu.M. *Instr. Exper. Techn.* **25** (1982) 778
9.6. Samosyuk V.N. and Fedoroff M. *Instr. Exper. Techn.* **33** (1990) 754

9.7. Kapitza S.P., Martynov Yu.T., Samosyuk V.N. *et al. Atomic Energy* **37** (1974) 356

9.8. Murashov M.V., Ph.D.Thesis, GIREDMET, Moscow, 1988

9.9. Segebade Ch., Weise H.P., Lutz G.I. Photon Activation Analysis, ed. W.de Gruiter, Berlin New-York, 1988

9.10. Kapitza S.P., Samosyuk V.N., Firsov V.I. *et al., J. Anal. Chem.* **39** (1984) 2101

9.11. Engelmann Ch., In: *Modern Analytical Techniques for Metals and Alloys*, ed. R.F.Bunshah, Ch.18, p.777, John Wiley and Sons, Inc., 1970

9.12. Holm D.M. and Sanders W.M. *Proc. Soc. Appl. Sprectr.*, June 1966

9.13. Persiani C., Spira J., Bastian R. *Talanta* **5** (1967) 565

9.14. Schweikert E. and Albert Ph. Radiochemical Methods of Analysis, IAEA, Vienna, 1965, p.323

9.15. Cuypers M. *Ann. Chem.* **13** (1964) 509

9.16. Kapitza S.P., Samosyuk V.N., Tsipenyuk Yu.M. *et al. Radiochem. Radioanal. Lett.* **5** (1970) 217

9.17. Chapyzhnikov B.A., Malikova E.V., Kunin L.L. *et al., J. Radioanal. Chem.* **17** (1973) 275

9.18. Wasserman A.V., Kunin L.L., Kapitza S.P. *et al., J. Anal. Chem.* **28** (1973) 729

9.19. Fridrich K., Mulbach H., Foigtman R. *et al., J. Anal. Chem.* **35** (1980) 682

9.20. Samosyuk V.N., Firsov V.I., Chapyzhnikov B.A. *et al., J. Radioanal. Chem.* **37** (1977) 203

9.21. Chapyzhnikov B.A., Evzhanov Kh., Malikova E.V. *et al., Radiochem. Radioanal. Lett.* **11** (1972) 275

9.22. Malikova E.V., Chapyzhnikov B.A., Kunin L.L. *Kernenergie* **5** (1978) 141

9.23. Chapyzhnikov B.A., Evzhanov Kh., Malikova E.V. *et al., Radiochem. Radioanal. Lett.* **11** (1972) 269

9.24. Revel G. *J. Radioanal. Chem.* **3** (1969) 421

9.25. Radionov V.I., Samosyuk V.N., Chapyzhnikov B.A. *et al. Radiochem. Radioanal. Lett.* **18** (1974) 379

9.26. Los-Nescovic C., Fedoroff M., Samosyuk V.N., Chapyzhnikov B.A. *Radiochem. Radioanal. Lett.* **45** (1980) 185

9.27. Fedoroff M., Los-Nescovis C., Samosyuk V.N. and Chapyzhnikov B.A. *J. Radioanal. Chem.* **72** (1982) 715

9.28. Fedoroff M., Los-Nescovis C., Rouchau J.-C. *et al. Erganzung zur 3 Taagung Nuklear Analysenverfahren* 1983 (Dresden) 9

9.29. Bock R. A Handbook of Decomposition Methods in Analytical Chemistry. Int. Textbook Co.

9.30. Fedoroff M., Los-Nescovis C., Revel G. *J. Radioanal. Chem.* **55** (1980) 219

9.31. Bodranski B. Analiza Ilosciowa Zwiazfow Organcznych. Warszaw, 1956

9.32. Fedoroff M., Los-Nescovis C., Rouchau J.-C. *et al., J. Radioanal. Nucl. Chem.* **45** (1985) 45

9.33. Kadik A.A., Matveev S.V., Tsipenyuk Yu.M. *et al.*, *J. Anal. Chem.* **49** (1994) 110

9.34. Samosyuk V.N., Firsov V.I., Chapyzhnikov .A. *et al. J. Radioanal. Chem.* **37** (1977) 203

9.35. Mazyukevich N.P., Shkoda-Ul'yanov V.A. *Atomic Energy* **21** (1966) 1069

9.36. Baranov V.I., Khristianov V.K., and Karasev B.V., *Isotopenpraxis* **3** (1967) 235

9.37. Overman R,F., Corey J.C., and Hawkins R.H., AEC Rep. DP-MS-68-9, Aiken, S.C., (1968)

9.38. Guinn V.P. and Lukens H.R. *Trans. Amer. Nucl. Soc.* **9** (1966) 106

9.39. Brezhnev O.N., Kazakevich G.M., Ponamarchuk V.A., and Fillipov E.M. *Instr. Exper. Techn.* **18** (1975) 338

9.40. Burmistenko Yu.N. Photonuclear Analysis of Substance Content (in Russian) Moscow, Energoatomizdat, 1986

9.41. Kapitza S.P., Martynov Yu.T., Sulin V.V., and Tsipenyuk Yu.M. *Isotopenpraxis* **12** (1976) 386

9.42. Leonard M., Ph.D.Thesis, Moscow, IPP, 1983

9.43. Martynov Yu.T., Sulin V.V., Tsipenyuk Yu.M. *Zavodskaya laboratoriya (in Russian)* **49** (1983) 51

9.44. Leonard M., Martynov Yu.T., Tsipenyuk Yu.M. *J. Anal. Chem.* **38** (1983) 1758

9.45. Belov A.G., Borcha E., Zhuchko V.E. *et al.* JINR 18-80-841, 1980

9.46. Burmistrov V.P. and Madiyanov T.N., *Atomic Energy* **40** (1976) 414

9.47. Fusben H.-U. and Segebade Chr. In: Development in Activation Analysis, Oxford, 1978, p.22

9.48. Ernandes A. and Zamyatnin Yu.S. *Isotopenpraxis* **20** (1984) 272

9.49. Ernandes R., Ph.D.Thesis, Dubna, JINR, 1984

Chapter 10

10.1. Burmistenko Yu.N. Photonuclear Analysis of Substance Content (in Russian). M.: Energoatomizdat, 1986

10.2. Hislop T.S. and Williams D.R. *J. Radioanal. Chem.* **16** (1973) 329

10.3. Engelman Ch. and Scherle A.C. *Int. J. Appl. Rad. Isot.* **22** (1972) 329

10.4. Engelman Ch., Filippi G., Gosset J and Moreau F. *Modern Trends 76*, München, FRG, v.1 (1976), p.644

10.5. McGregor M.H. *Nucleonics* **15** (1957) 176

10.6. Malinin A.B. *et al.*, *Radiochemistry* **12** (1970) 780

10.7. Marceau M. *et al.*, *Int. J. Appl. Radiation Isotopes* **21** (1970) 667

10.8. Kato T. *et al.*, *Talanta* **19** (1972) 515

10.9. Yaqi M. *et al.*, *Chem. Letters* **3** (1972) 215

10.10. Gray F.C. *et al.*, *J. Nucl. Med.* **14** (1973) 931

10.11. Levin V.I., Malinin A.B., Tronova I.N. *Radiochem. Radioanal. Lett.* **49** (1981) 111

10.12. O'Mara R.E. *at al.*, *J. Nucl. Med.* **9** (1968) 340

10.13. Hisada K. *et al.*, *J. Nucl. Med.* **14** (1973) 615

10.14. Tronova I.N. *et al.*, *Medical Radiology)* **8** (1973) 31

10.15. Chandra R. *et al.* *Radiology* **100** (1971) 687

10.16. Levin V.I. *et al.*, *Radiochemistry)* **19** (1977) 388

10.17. Goodwin D.A., Finston R.A., Colombetti L.G. *et al.* *J. Nucl. Med.* **10** (1969) 337

10.18. Bochkarev V.V. Levin V.I., Stanko V.I. *et al.*, *Medical Radiology* **1** (1974) 4

10.19. Neirinckx R.D. *Radiochem. Radioanal. Lett.* **4** (1970) 153

10.20. Thakur M.L., Nunn A.D. *Int. J. Appl. Rad. Isot.* **23** (1972) 139

10.21. Levin V.I., Kozlova M.D., Malinin A.B. *et al.* *Int. J. Appl. Rad. Isot.* **25** (1974) 286

10.22. Dahe J., Tilbury R. *Int. J. Appl. Rad. Isot.* **23** (1972) 431

10.23. MacDonald N.S., Neely H.H., Wood R.A. *et al.* *Int. J. Appl. Rad. Isot.* **26** (1975) 631

10.24. Brown L., Beets A. *Int. J. Appl. Rad. Isot.* **23** (1972) 57

10.25. Malinin A., Kurenkov N., Kozlova M., Sevastyanova A. *Radiochem. Radioanal. Lett.* **59** (1983) 213

10.26. Davydov M.G., Mareskin S.A. *Radiochemistry* No.5 (1993) 91

10.27. Davydov M.G., Shcherbachenko V.A. *Atomic Energy* **27** (1969) 205

10.28. Nordell B.O., Wagenbach U., Sattler E.L. *Int. J. Appl. Rad. Isot.* **33** (19820 183

10.29. Malinin A.B., Kurenkov N.V. *Radiochem. Radioanal. Lett.* **53** (1982) 311

10.30. Oganessyan Yu.Ts., Starodub G.Ya., Buklanov G.V. *et al.* *Atomic Energy* **68** (1990) 271

10.31. Oka Y., Kato T., Nomura K., Saito T. *Bull. Chem. Soc. Jap.* **40** (1967) 573

Chapter 11

11.1. Egawa S., Tsukiyama I., Ono R. *et al.*, *Jpn. J. Clin. Oncol.* **14** (1984) 613

11.2. Wideroe R. *Proc. V Int. Betatron Symp.*, Bucharest, 1971, p.397

11.3. Barendsen G.W. *Response of Cultured Cells, Tumours and Normal Tissues to Radiations of Different LET*. In: Current Topics in Radiation Research, eds. M. Ebert and A. Howard, North-Holland, Amsterdam, 1968, v.4, p.295

11.4. Rosander S. *'Microtrons as Therapeutic Tools'*, Indo-Soviet Symposium on Microtrons, CAT, Indore (India), 1993

11.5. Kozlov A.P. and Shishlov V.A. *Acta Radiologica* **15** (1976) 493

11.6. International Electrotechnical Commission. Publication 601-2-1. Safety of medical electrical equipment, IEC, Geneva, 1981

11.7. Brahme A., Kraepelien T and Svensson H. *Acta Radiologica Oncology* **19** (1980) 305

11.8. MM50 Advanced Radiotherapy Treatment System, *Scanditronix Medical* 1996

11.9. Karlsson M., Nyström H., and Svensson H. *Med. Phys.* **19** (1992) 307

11.10. Karlsson M., Nyström H., and Svensson H. *Med. Phys.* **20** (1993) 143

Chapter 12

12.1. Balasko M. and Svab E. *Nucl. Instr. Meth.* **A 377** (1996) 140

12.2. Zakirov B.S., Kapitza S.P., B.I.Leonov *et al. Defectoscopy* **1** (1987) 32

12.3. Luk'yanenko E.A., Gusev E.A., Leonov B.I., and Zakirov B.S. *Instr. Exp. Techn.* **19** (1976) 988

12.4. Rosander S. and Trower W.P. *Report in Indo-Soviet Seminar on Microtrons*, CAT, Indore (India) 1992

12.5. Trower W.P. *Nucl. Instr. Meth.* **B 79** (1993) 589

12.6. Oka Y., Kato T., Nomura K. and Saito T. *Bull. Chem. Soc. Jap.* **40** (1967) 575

12.7. Řanda Z., Ducháček V., Hradil M. *Jaderna Energie* **34** (1988) 365

12.8. Gerbish Sh., Sodnom N., Baatarkhuu D., Zhuchko V.E., Leonard M., Peres G., Riige, Communication of JINR, P6-91-123, Dubna, 1991

12.9. Galatanu V., Grecescu M. *Rev. Roum. Phys.* **24** (1979) 9

12.10. Kato T. and Oka Y. *Talanta* **19** (1972) 515

12.11. Karlsson M. *Report ISSN 1103-6605*, Royal Inst. Techn., Stockholm, 1993

12.12. Bizzarri U., Giubileo G., Messina G. *et al. Nucl. Instr. Meth.* **B 50** (1990) 331

Index

T - #0034 - 111024 - C0 - 229/152/21 - PB - 9780367396862 - Gloss Lamination